Technical Communication and Its Applications

Technical Communication and Its Applications

Jerome N. Borowick
California Polytechnic University - Pomona

Prentice Hall
Englewood Cliffs, New Jersey   Columbus, Ohio

Library of Congress Cataloging-in-Publication Data

Borowick, Jerome N.
    Technical communication and its applications / Jerome N. Borowick.
      p.   cm.
    Includes index.
    ISBN 0-13-199275-9
    1. Technical writing.   2. Communication of technical information.
I. Title.
T11.B66   1996
808'.0666—dc20   95-35883
                                                                  CIP

Editor: Stephen Helba
Production Editor: Mary M. Irvin
Design Coordinator: Julia Z. Van Hook
Text Designer: Rebecca Bobb
Cover Designer: Tracey Ward
Product Manager: Debbie Yarnell

This book was set in Century Schoolbook by Carlisle Communications, Inc., and was printed and bound by Book Press, Inc., a Quebecor America Book Group Company. The cover was printed by Phoenix Color Corp.

 © 1996 by Prentice-Hall, Inc.
A Simon & Schuster Company
Englewood Cliffs, New Jersey 07632

Printed in the United States of America

10 9 8 7 6 5 4 3 2 1

ISBN: 0-13-199275-9

Prentice-Hall International (UK) Limited, *London*
Prentice-Hall of Australia Pty. Limited, *Sydney*
Prentice-Hall of Canada, Inc., *Toronto*
Prentice-Hall Hispanoamericana, S. A., *Mexico*
Prentice-Hall of India Private Limited, *New Delhi*
Prentice-Hall of Japan, Inc., *Tokyo*
Simon & Schuster Asia Pte. Ltd., *Singapore*
Editora Prentice-Hall do Brasil, Ltda., *Rio de Janeiro*

Preface

Successful technologists, engineers, and scientists must have effective communication skills. To develop these skills, technical institutes, colleges, and universities have varying approaches: some require a series of writing or communication classes that are completed throughout the student's education; many require only one or two writing classes that may be completed during the freshman year. These classes usually emphasize grammar, sentence structure, and syntax for effective communication but—for the specific reports and documents that are typically written by technologists, engineers, and scientists—a sense of specific audience, a detailed understanding of purpose, and the protocol of the technical documentation process are not strongly emphasized.

Laboratory classes address one important category of technical documents; however, emphasis is usually on the laboratory procedure, data, and analysis rather than on effective communication skills. Also, many schools require students to complete at least one public speaking class and may require oral presentations as a component of their technical and general education classes. The importance of making effective oral presentations cannot be overemphasized.

As a student, I was required to write many papers for my general education classes. However, the skills that I developed from these assignments did not prepare me for the writing requirements of my profession.

I realized shortly after I graduated from college that communication skills are important to make others aware of the results of my work. After I developed these skills, I found that I was always given the appropriate recognition for my technical accomplishments by my supervisors and clients and that my communication skills contributed to my success as a technical professional more than any other skill.

This text bridges the gap between academia and the professional world so that young professionals will experience a smooth transition to their first professional employment after graduation. It builds on the communication skills and knowledge that students presently have and teaches the elements of the various types of engineering and scientific documents and oral presentations that a technical professional is expected to be familiar with.

METHOD OF INSTRUCTION USED IN THIS TEXT

This text explains the principles of technical writing as used by the practicing technical professional in industry, government, and academia.

Each of the four sections of this text has a distinct purpose:

- Section I emphasizes the principles of technical writing for clarity and effectiveness.
- Section II teaches the framework for writing technical reports and includes simulated components of professional reports.
- Section III introduces the various types of reports and documents written by students and practicing professionals. This section includes annotated examples that reveal frequent student problems and includes simulated components of professional reports and examples that emphasize the concepts addressed in this text. Therefore, it is important to carefully study these simulated components and examples and to adapt the structure, content, and style to meet specific needs.
- Section IV discusses personal communication of the professional.

The hypothetical exercises at the end of each chapter realistically reflect professional practice with one modification: Because typical undergraduate students have limited technical and scientific knowledge to draw on, the exercises relate to the familiar subjects, projects, and activities of students, and may include campus life and students activities. After discussion with your instructor, the facts and data presented in each exercise may be supplemented with purposes and contexts that may limit the scope of the report to either already familiar or readily available information and knowledge.

The headings and sections presented in this text are meant to be guidelines for the structuring of typical technical reports. Professionals, however, often need to modify headings and sections to meet the needs of their audience and the purpose of their specific reports. Students are encouraged to do the same.

As a word of encouragement, writing, like any other skill, becomes easier with practice. At first you may find the assignments difficult and time-consuming. With practice you will begin to understand the technical writing method and become more proficient at expressing yourself.

I hope you will find, as I did, that technical and scientific communication can be a very enjoyable and rewarding experience.

I thank the many hundreds of students and the Civil Engineering faculty at Cal Poly, Pomona, who have since 1981 encouraged and helped me to write this text. I also acknowledge the many companies that have submitted reports and documents to me for my review and use and the many students who have graciously permitted me to use their class submittals in this text. I also thank Hovel Babikian for his assistance in completing Chapter 11, " Student Laboratory Reports," and for his guidance concerning computer technology. I want to thank all of the reviewers of this book for their careful consideration and comments: Marian M. Clark, East Tennessee State University; Mary E. Debs, University of Cincinnati; Richard M. Drake, Fluor Daniel, Inc.; Marilyn Dyrud, Oregon Institute of Technology; Harold Erickson, Duluth Technical College; Lory Hawkes; Donald G. King, Donald G. King Associates; Joanne H. Kulachok, BSI Consultants, Inc.; Ann McGuire, Dunwoody Institute; Stephen O'Neill; and Tom Zimanzl, Roosevelt University-Robin Campus.

I thank my son, Kent, and my daughter, Tami, for their encouragement and advice regarding content and pedagogy. And, last but not least, I thank my wife, Carol, for her encouragement to write this text, guidance, proofreading, and enduring my many years of early morning work habits.

Jerome N. Borowick

Contents

Elements of Technical Writing

Section I introduces the purposes and principles of technical writing. These provide an understanding of the technical writing process and create the basis for writing technical reports.

Chapter 1, "An Introduction to Technical Writing," discusses the importance and style of communicating technical information. To provide an understanding of the applications of technical writing, the purposes and contents of technical reports are introduced. And the needs of the audiences for which these technical reports are intended are discussed.

The step-by-step procedure for effective technical writing is presented in Chapter 2, "The Technical Writing Process." Also, writer's block is addressed, and advice to non-native writers is included in this chapter.

Chapter 3, "Principles of Clear Technical Writing," and Chapter 4, "Rules of Practice for Technical Writing," discuss the principles of communicating technical information effectively to your audiences.

After studying Section I, you will be prepared to learn to write the components of technical reports.

CHAPTER 1

An Introduction to Technical Writing

Technical writing is the critical link between professionals.

THE DEFINITION OF TECHNICAL WRITING

Technical writing is the written communication of engineering and scientific ideas, concepts, and data presented objectively, logically, and accurately. The writing is clear and concise. Graphics are usually included to assist the reader in understanding the subject matter. The reader should be convinced of the conclusions based on the information presented. The recipient will usually use the report or document to perform a task, make a decision, solve a problem, or acquire information and knowledge.

THE IMPORTANCE OF TECHNICAL WRITING

As a technical professional, you will have the opportunity to perform many functions: research, design, analyze, manufacture or construct, test, and manage. In the performance of these functions, many technologists, engineers, and scientists new to the work environment discover that they may spend as much as 50 percent of their time writing reports and documents that discuss the results of their work.

Your technical competence will be important to the successful completion of the projects you work on. In today's high-tech industrial society, it is likely that your work on these projects will be communicated and coordinated with other professionals, government agencies, clients, and managers who depend on your results. Therefore, the successful completion and profits of any project may be jeopardized without effective communication between professionals that discusses progress and problems.

A few years ago, the Chief Executive Officer of Lockheed-California said the following to a group of engineering educators:

From a purely technical and scientific standpoint, your engineering colleges are sending us the best-educated, best-prepared new engineering graduates we have ever had. We are strained to challenge their technical and quantitative abilities. However, these same graduates have one glaring deficiency: they can't write. I want you to send me engineers who can WRITE! WRITE! WRITE! WRITE! WRITE!

Messages similar to this one are repeated over and over by leaders of industry, government, and academia. The importance of your ability to technically write well cannot be understated.

From a personal standpoint, your peers and colleagues sometimes will judge your technical competence by evaluating the effectiveness of your writing. Frequently, performance reviews by supervisors are subtle reflections of your ability to write. Your reports and documents may be your only form of communication with clients, government agencies, and professionals at other facilities and companies. Your reputation as a professional depends not only on your ability to perform well, but also on your ability to write well.

THE STYLE OF TECHNICAL WRITING

Effective technical writing is objective, clear, concise, and convincing. This is accomplished when the style is descriptive and quantitative; that is, when the writing includes details and uses facts, data, measurements, and statistics. All pertinent information is presented in a manner that can be objectively evaluated and concluded by the reader. Imprecise, judgmental, emotional, and editorial words such as *many*, *undue*, or *annoying* are avoided to deter readers from interpreting the magnitude of the meanings of these words based on their own personal experiences. Only information that is relevant to the purposes of the reports is included. Information that is interesting, but not pertinent, is excluded. The purpose of the communication determines the tone (manner and attitude of expression).

College writing addresses an audience that knows more than the writer and that needs a basis for determining the level of understanding of the writer (student). This writing discusses already proven concepts and ideas supported by selected facts and data. By contrast, technical writing addresses an audience that knows less than the writer and that needs information. This writing discusses unknown concepts and ideas supported by all pertinent facts and data. Therefore, the information presented in technical writing must be persuasive by including facts, data, and analysis rather than by convincing argument.

LISTS USED IN THIS TEXT

Lists, when used to present information, are effective for organizing your material into a series of discrete items for discussion and for helping readers to comprehend and remember the information presented.

- When the items in a list are preceded by numbers or letters in sequence (e.g., 1, 2, etc., or a, b, etc.), the sequence of operation is important and should be followed in the order presented. This convention is typical for procedures and other chronological series of events.

- When the items in a list are preceded by bullets, (such as in this list) the items are unrelated and can be addressed in any sequence. However, the items are usually included in order of descending importance.

DEFINITIONS OF TERMS USED IN THIS TEXT

For clarity, the following definitions of terms used in this text are provided:

- **Professional** includes engineer, scientist, and other technical professional.
- **Report** includes report and document.
- **Reader** includes any member of your intended audience.
- **Heading** includes topical heading of a section that discusses subject matter (e.g., Design Parameters, Test Equipment) and functional heading of a section that discusses the function of the subject matter (e.g., abstract, summary).
- **Section** includes the paragraphs where the topical discussion of subject matter (e.g., Design Parameters, Test Equipment) and functional discussion of the subject matter (e.g., abstract, summary) are found.

TYPES OF TECHNICAL REPORTS

As a professional, you will be expected to write many communications and reports to serve different purposes and audiences. Although these communications and reports may have different generic titles, they will fall into the following categories, which are discussed in detail in Section III.

Business Communications

Business communications transfer information concerning the management and administration of products, projects, and personnel. See Figure 1–1.

- **Business letters, memos, and e-mail** are the foundation for conveying information in business. They are the quickest and most common form of communication and are used for purposes such as requests for information and action.
- **Periodic, progress, and trip reports** communicate progress on projects and events to those concerned with their completion and success. Readers review, and may revise, scheduled courses of action as a result of the information included in these reports.
- **Personnel evaluations** review the past performance of employees to foster better working relationships between employees and employers.

FIGURE 1–1
Types of Business
Communications

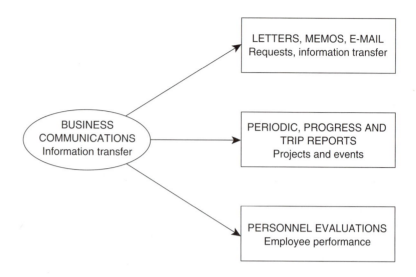

Planning Reports

Planning reports determine and control future events. See Figure 1–2.

- **Activity and product evaluations** investigate past or present activities and products to determine future courses of action.
- **Bids and proposals** suggest a course of action at a specific price to a potential client to obtain a contract. Bids and proposals are usually submitted in response to a request from the potential client.
- **Specifications** contractually communicate the needs of a client to a supplier of goods or a provider of services. They are the requirements with which the supplier or provider must comply.
- **Feasibility reports** investigate the economic and technical practicality of a proposed course of action.
- **Environmental impact reports** evaluate the effects of a proposed project on the environment and are written to comply with government regulations. Environmental impact reports are also used for planning.

Deterministic Reports

Deterministic reports (reports that discuss occurrences determined by previous events) study and evaluate existing facts, data, and concepts (see Figure 1–3).

- **Technical articles** disseminate technical information to interested professionals and colleagues. An acknowledgment or response is rarely requested from the reader. Technical magazines and journals are usually used for this purpose.

FIGURE 1–2
Types of Planning Reports

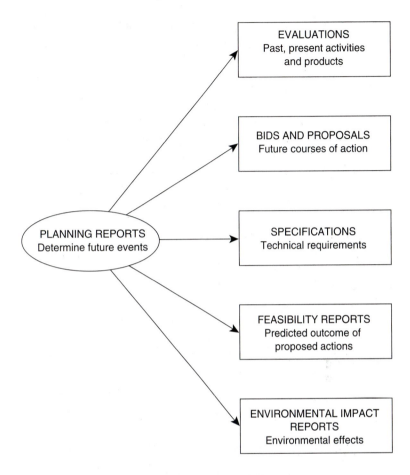

- **Laboratory reports** document laboratory procedures and their results. Laboratory reports may be one component of a research project or may demonstrate technical capability and compliance with contracts.
- **Scientific and engineering analyses** are mathematical evaluations of scientific and technical systems. Because ideal conditions are usually assumed, the results of analyses are usually less reliable than the results of tests. However, analyses are very cost-effective. Scientific and engineering analyses are usually essential components of scientific and engineering projects.
- **Research reports** communicate the results of scientific and sociotechnologic investigations to help understand scientific and sociotechnologic phenomena. These reports may include analytical as well as laboratory components.

AUDIENCES

Understanding the perspective of your audience is the key to writing effectively. In literary and journalistic writing, because of the uncertain appeal to readers of

FIGURE 1–3
Types of Deterministic Reports

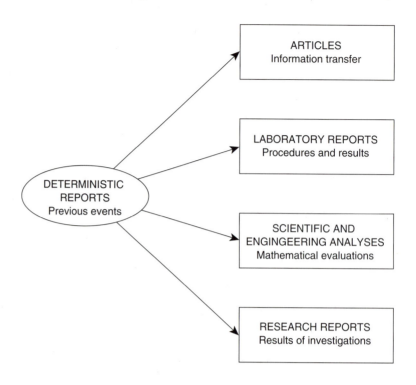

writing intended for entertainment, the audience often consists of unidentified and divergent members. This makes the interests and background of the members of a literary audience difficult to analyze.

On the contrary, the audience for technical reports is ordinarily well defined because you are writing for a specific purpose and can direct your reports to those with a specific capability to help you achieve your goal or to those with an interest in your results. Frequently, you or someone in your organization may have personal knowledge of your readers. These advantages simplify the identification of the members of your audience.

Technical readers ordinarily have a professional or organizational responsibility to read the material intended for them and have the ability to comprehend it. However, unlike literary and journalistic readers, these readers may not be enthusiastic concerning the material, which usually requires an active response, and they can choose to either respond to or ignore it. Unless the writer appeals to their concerns, these readers will usually ignore the material. Therefore, understanding the needs of your readers is essential to make your writing effective.

The readers of your reports will fall into one or more of the following categories (also see Figure 1–4):

- **Executives** Executives need to make decisions based on applicability, marketability, and profitability. They want conclusions and alternatives rather than details.

FIGURE 1–4
Technical Report Audiences

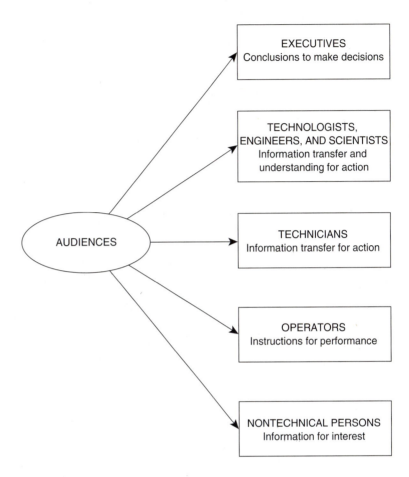

- **Technologists, Engineers, and Scientists** Technologists, engineers, and scientists are interested in information transfer. They need facts, details, theory, methodology, and conclusions that are substantiated by the information presented in the report. These readers sometimes refer to other sources of information referenced in the report.

- **Technicians** Technicians need information to troubleshoot, modify or upgrade, and maintain or repair equipment. They need practical information in a format that is easy to use. To facilitate understanding of subject matter, technicians rely on visuals such as tables, graphs, drawings, and photographs to supplement the written words.

- **Operators** Operators need instructions to operate equipment or to perform procedures. A set of easy-to-understand commands in a step-by-step format with visuals, where necessary, prevents misunderstanding the intent.

- **Nontechnical Persons** Nontechnical persons read for interest and information. Their interest may be casual, such as in articles found in periodicals, or it may be to understand technology that may affect social issues, such as in reports to a city council concerning the environmental impact of a proposed project or to an attorney concerning liability for a product deficiency. These readers are most interested in definitions, descriptions, causes and effects, and applications. Photographs and charts enhance their understanding of the material.

WRITING FOR MULTIPLE AUDIENCES

It is simpler to write a report that addresses a single audience (e.g., executives or operators) than a report that addresses more than one audience (e.g., executives and operators). For a report intended to be read by a multiple audience, you can use one of the following techniques:

- Precede all information with headings that direct different readers to sections of the report relevant to them. The words used in these headings must accurately represent the content of the included information. Information can be repeated in the different sections of the report from the perspective of the audience intended for those sections.
- For each audience, write a different cover letter that emphasizes the relevant sections of the report and adds any other relevant information.
- Write a separate, similar report for each audience. A word processor simplifies revising the document for each audience.

Most reports that address multiple audiences use the first method because it is time effective, generates less paper, and permits all readers easy access to additional information, when desired.

TEAM REPORTS

Frequently, reports are written as a concerted effort of a team of several professionals working in different specialties or disciplines. This experience can be very rewarding and challenging because it is an opportunity to learn from, and contribute to, the expertise of the other professionals in the team.

Writers have individual perspectives regarding the organization and style of writing for the report. A team leader, either appointed by management of the organization or chosen by the other writers before the writing begins, can define the scope of responsibility for each professional in the team to avoid duplication of effort. The team leader can also establish the quality of workmanship and standard of presentation.

The team leader needs to organize and coordinate the writing of the report and to direct the efforts of the writers toward preparing a consistent, well-written report. Each writer is responsible for cooperating with the other writers and learning from the leadership and judgment of the team leader.

Unfortunately, special problems such as irreconcilable viewpoints between members, shirking of responsibilities, and intervention of organizational politics sometimes arise when writing team reports. These problems need to be resolved tactfully by the team leader in a manner that is workable and deemed acceptable by all team members. Occasionally, management personnel resolve these problems and may relieve some team members of their responsibilities. When this action occurs, it should not be perceived as a win or loss by individual team members, but rather as an action intended to benefit the organization and, therefore, all members of it. In spite of its potential drawbacks, writing team reports is frequently the most expedient method to achieve an organizational objective and should be an experience to be eagerly anticipated.

WORD PROCESSORS AND WORD PROCESSING PROGRAMS

The writing process has been revolutionized by word processors and word processing programs. Writers can easily and quickly revise, edit, and check spelling with a spell-checking program. Headings can be enlarged and differentiated from the body of the text with different fonts and styles. Graphics can be inserted into the body of the text. The time required to write a report is reduced considerably. First drafts can use the same structure and format as the intended final drafts. As a result, writing has become more gratifying for many professionals, and the quality of their work has improved. You are encouraged to learn to use a word processor or word processing program as soon as possible.

Computer programs are available to check grammar and readability of writing. These programs can help you improve your writing, but they are merely tools; the responsibility for learning to write is yours. Spell-checking programs are excellent for eliminating spelling errors, but they cannot differentiate between correctly spelled homonyms that have unrelated meanings for the intended words. For example, "The break must be repaired" is concerned with a structural deficiency, whereas "The brake must be repaired" is concerned with a mechanical deficiency. Because these homonyms can mislead the reader and result in an unintended message, the writer still needs to proofread carefully.

KEY CONCEPTS

- Technical writing is written with a purpose and for a specific audience. It is not intended to entertain the reader.
- The contents of technical reports are determined by the purposes for which they are written and the intended audiences.
- The tone (manner and attitude of expression) of technical reports is determined by the purposes for which these reports are written and by their intended audiences.

- Letters and memos are the most common form of business communication. Ordinarily, information included in them is timely, and readers give this form of communication immediate attention.

- Technical reports written for multiple audiences require careful selection of the titles for headings to direct readers to the applicable information.

- Team reports provide opportunities to expand your knowledge as well as contribute to other disciplines. These reports also help you learn to work in small groups—the structure of the professional work environment.

- Reports written with word processors or word processor programs can enhance the quality of your reports.

STUDENT ASSIGNMENT

Find one of each of the following (possible sources include your school library, periodicals and technical journals, your place of employment, and your daily correspondence), and determine the type of technical report, its purpose, and audience:

1. A business communication.
2. A planning report.
3. A deterministic report.

Note: The results of this assignment will be used for the Student Assignment in Chapter 2.

The Technical Writing Process

*Defining your purpose and understanding your audience are
the keys to effective technical writing.*

Effective writing is time-consuming. Inexperienced professionals commonly err by allowing inadequate time to prepare their reports properly. The technical writing process should not be a concentrated effort; rather, it should be several smaller efforts separated in time to help you organize your ideas. It is most efficient to begin writing the components of a report as you complete the phases of your work project so that when you are ready to write the report, parts of it may be ready for rewriting and editing for the final draft.

Technical writing, more than literary and journalistic writing, is a recursive process. As components of a report are completed, information presented in earlier sections may need to be supplemented, revised, or deleted so that these components become the natural results of this information. This recursiveness may occur during any stage of the writing process discussed in this chapter.

THE WRITING PROCESS IN SIX STAGES

The six-stage process described next (see Figure 2–1) is recommended for writing your report:

1. Define your problem and purpose.   What problem needs to be resolved; what do you intend to accomplish?

- Do you want the reader to follow a course of action? When the report is intended to persuade the reader, recommendations must be expressed in practical terms.
- Does the reader need information? Facts and data must be clear and discussion built on information previously presented. Conclusions must be substantiated with the information included in the report.

FIGURE 2–1
The Technical Writing Process

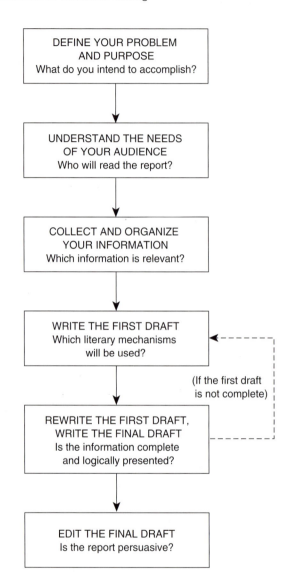

2. **Understand the needs of your audience.**   Who will read the report?

- Technical people? An explanation of technical principles should be neither too simplistic nor too complex for the level of the reader's background.

- Nontechnical people? The content should appeal to their purpose for reading the report. For example, a report to the planning commission should emphasize the benefits of a proposed project to the general public. Technical principles, when included, should be explained as simply as possible with analogies and graphics.

- Managers? Recommendations must include sufficient information for easy implementation. Cost-effectiveness and efficiency are usually addressed.

Technical content, when included, can be conclusive rather than descriptive or analytical.

- Persons with less education than the writer? Sentences should be shorter and the vocabulary simpler than when writing for a peer audience. Technical concepts can be simply explained by analogy using the principles of high school physics and chemistry.

- Persons with a different environmental, geographical, or cultural background? Analogies, comparisons, and examples should be applicable to their understanding. For example, analogize cold by discussing ice cubes rather than icicles for an audience familiar with a warm climate. Jargon and slang should always be avoided; examples should not be culturally offensive.

- Multiple classifications of readers? This is the most difficult audience to write for. For most writers, this is also the most frequent audience. Sufficient information should be included to satisfy the needs of all readers. Information should be organized into sections and identified with headings for readers to select the sections that pertain to their interests. The presentation of the information in each section should not assume knowledge of information presented in prior sections. Therefore, information from prior sections should be referenced as necessary or may be presented again from a different perspective.

Also, consider the following:

- What is your reader's motivation for accepting the findings in your report? When the reader needs to be persuaded, such as in a proposal, attention must be given to attractive, as well as functional, graphics and layout. On the other hand, an instruction manual must provide sufficient details and graphics for performing the task and does not need to be attractive.

- What is the professional experience level of the readers? The explanation for an electrical circuit failure must be more general to engineering managers than to design engineers. However, engineering managers want details concerning the cost and schedule for repair.

- What is the organizational role of your readers? City planners, although frequently technically aware, are more concerned with benefits, environmental effects, and code compliance than technical feasibility.

3. Collect and organize your information.   You will spend considerable effort gathering facts, data, and other important information required for your project or task.

  a. Collect all the information that you have used or generated (e.g., notes, data, calculations) to complete the project or task.

  b. Recognize that in completing a project or task, the procedure for successful completion ordinarily is not well defined, and therefore, much of the collected information may be the result of misdirected or unsuccessful trial-and-error attempts. Determine which information is relevant to the purpose of the report and should be included. Place nonrelevant information in a file for possible future use. For example, while collecting data in a research and development project intended to determine a process for improving the mechanical properties of a material, it is experimentally determined that exposure of this material to pure oxygen during annealing has no significant beneficial effect. This information should be included in the research and development report for other potential applications. However, this same information should not be included in a specification that defines the manufacturing process.

c. Separate information into the following four categories: statement of the problem; additional information used to solve the problem; work performed to obtain an outcome; and the outcome. As a guideline, refer to Section III, "Applications for Students and Professionals." Use the chapter that discusses your category of report to select sections for including in the report. Separate the information within each of the four stated categories (i.e., statement of the problem, etc.) into these sections. Many writers use file folders or loose-leaf binders with separators for this purpose.

d. Use the same chapter of Section III as a guideline to determine the most suitable sequence for presentation of your information in these sections of the report. Arrange your sections into this sequence. Then, prepare an outline of information to be included in each section of the report.

e. Establish a reasonable time frame for writing your report. Consider that, even though actual writing time may be only hours, a well-written report may require days or weeks to complete.

4. Write the first draft.   Several different literary mechanisms can be used effectively to impart the different aspects of information presented in the report. Each section may use one, or a combination, of the following literary mechanisms:

- Descriptions create visual images, and explanations interpret occurrences and phenomena. Frequently, an explanation will accompany a description. Events, objects, physical concepts, and outcomes are ordinarily described and explained.

- Chronologies present a series of events. Occurrences, historical backgrounds, process descriptions, and instructions and procedures are ordinarily presented chronologically. Many descriptions and explanations are presented chronologically.

- Analyses, calculations, test data, and results evaluate the given information and data. The sections of the report that include these topics determine the outcomes for the problem.

- Examples explain concepts and theory. Introductory and descriptive sections of reports frequently include examples. Recommendations and alternatives may also include examples.

- Analogies relate the similarities between the unknown and the known and are frequently included in the descriptive and explanatory sections of research reports. For example, the size and shape of an object unfamiliar to the reader can be analogized to a household item such as a toaster or couch.

- Comparisons are similar to analogies except that they relate the differences between the unknown and known. For example, the odor of an object unfamiliar to the reader can be compared to the odor of gasoline, but sweeter.

- Graphics facilitate and reinforce understanding the text material. Also, they help the reader to visualize ideas, concepts, and facts, and remember them.

Frequently, sections of the first draft are written as the project progresses. If you have not done this, begin by writing the easiest section first. Many writers begin with the body of the report. Data reduction or numerical analysis is usually a good starting place. Technical descriptions and procedures are also good starting places. The abstract and summary are usually the last sections to be written because these sections require that you understand the contents of the report.

If you are having difficulty writing one or more of the sections (see the next section, "Coping With Writer's Block") you may begin Stage 5 before the first draft of all sections is completed. Writer's block is a common problem even for experienced writers. However, remember that you must eventually return to Stage 4 to complete those sections.

Before you proceed to Stage 5, you should go on to other unrelated tasks for several days. This will give you a fresh perspective, and you will be able to review and rewrite your draft in a more objective manner.

5. Rewrite the first draft, write the final draft.   Review whether the information is easily understood as it is introduced, and check its completeness for appropriate comprehension.

- First drafts frequently assume too much knowledge and background of the reader. Is more explanation required? Sometimes you may realize that you have omitted a key concept, and you may need to add an entire paragraph; other times you may need to add only a modifier.

- Is information repeated unnecessarily? Is irrelevant information included?

Review the ease of comprehension, accomplishment of purpose, and accuracy. Reading the first draft aloud will help identify slips in clarity and punctuation errors.

- Is your writing concise and to the point? Are your style and tone appropriate?

- Have acronyms been defined? Will the reader understand your abbreviations?

- Are the grammar, spelling, and calculations correct?

- Have you avoided clichés and slang?

Revise your draft until you are confident that the reader can understand your report without further explanation. Most reports are revised several times before the final draft is edited.

6. Edit the final draft.   The presentation of your report should be impeccable when it is submitted to the reader.

Structure and format:

- Does the title page, when included, have the necessary information, and is it professional looking?
- When required, does the report follow a standard format requested by the recipient?
- Are the margins, headers, footers, and headings logically and esthetically balanced with each other and the text material?
- Are the pages numbered and placed in the proper sequence?
- Do the titles of the visuals represent their content? Are the visuals referenced in the text of the report?

Accuracy and legibility:

- Are there any grammatical, spelling, numerical, or typographical errors?
- Is the print dark and large enough to read easily?

The final editing is usually performed by the writer, but it is a good idea to have it proofread for clarity by a co-worker not familiar with the project.

COPING WITH WRITER'S BLOCK

Writers frequently have difficulty organizing their thoughts and, therefore, beginning writing. This occurrence, commonly known as writer's block, is temporary and should not be attributed to an inability to write. Several different techniques are effective for writers working through writer's block. The most commonly used follow:

- Using a thesaurus, work through a complex concept by finding the appropriate words that express your ideas. Then, organize these words into phrases, these phrases into sentences, and these sentences into paragraphs. Revise the passage until it clearly expresses your thoughts.
- Review the parts of the report that are written to determine the unfinished subject matter that needs to be addressed.
- Write other sections of the report until their completion clarifies the purpose of the unfinished section. Then, return to the unfinished section.
- Obtain and read other relevant material to give you new ideas.
- Request the help of a colleague or associate for a fresh perspective. Avoid making judgments, and carefully consider any suggestions.
- Brainstorm with other professionals familiar with your project. Consider any approaches that have been successful in the past.

ADVICE FOR NON-NATIVE WRITERS

Students whose first language is other than English frequently have anxieties about technical writing. If you are one of these students (or even if you are an experienced writer), consider that all technical writing requires the following procedures:

> Determine the problem to be solved.
>
> Collect information and data.
>
> Organize the report.
>
> Analyze the information and data.
>
> Develop visuals.
>
> Write the text of the report.
>
> Proofread.
>
> Establish the format and visual design.

Non-native writers are usually as capable as writers in all these procedures except, because of their bilingual background, writing and proofreading the text of the report. This difference in background can be overcome by following this procedure:

1. Determine the problem to be solved and the appropriate method for your audience that will be used for its solution.

2. Carefully select the useful information and data for inclusion in your report. Do not include any nonrelevant information and data in your report, because it diverts readers from your intended objective.

3. Organize the structure of your report and select headings so that readers will be required to read the text material to obtain the details only.

4. Fully develop your ideas in the text material, but express them concisely to reduce the number of writing errors.

    Use a language dictionary to help you translate your thoughts into English. Use a thesaurus to help you select the appropriate words to express your ideas. When you are uncertain of the proper use of a word, check its literal meaning in an English dictionary. Ask an experienced writer to help you select the appropriate nouns for technical configurations and concepts (e.g., slotted hole, tongue-and-groove, over-the-center).

5. Support the written text material with calculations and visuals whenever possible. Carefully introduce and label all nontext material.

6. When you have completed your report to the best of your ability, have it proofread by an experienced writer. Carefully study the revisions, and review them with the proofreader.

7. The writing difficulties of non-native writers are typically caused by the improper use of words and idioms and the inability to apply the rules of English grammar appropriately, even though these rules may be understood.

Several days after the final draft is completed, you can use this understanding to your advantage by proofreading your own text material to study your application of these rules.

Most schools have a learning resource center or other tutorial service for students who need special assistance with their writing. The cost for services is usually minimal and sometimes may be free. As a student, take advantage of the opportunity to use these services whenever your classes require writing.

Writing improves with practice for all writers. Give yourself the opportunity to write whenever you can. Write letters to friends and relatives, keep a diary, and take writing classes. Have your writing assessed by a writing professional whenever possible. With practice, you will gain confidence, and your ability to write effective technical reports will increase.

KEY CONCEPTS

- Effective technical writing is time-consuming. Sufficient time should be scheduled to write your report. Preparing the first draft of your report as the project progresses expedites the writing.
- An effective writer understands the purpose and audience of the report.
- It is important to organize your material before you begin writing.
- Selecting the most effective literary mechanism helps the reader comprehend your information.
- Writer's block occurs for all writers. It is overcome easily when you realize that writer's block is temporary. Do not attribute its occurrence to your inability to write.
- A co-worker should proofread your final draft for clarity and ease of comprehension.

STUDENT ASSIGNMENT

For each of the technical reports used for the Student Assignment in Chapter 1:

1. Determine the possible sources of the information.
2. Discuss the relevancy of the information to the purpose of the report. Is any information not included that should be? Is any information included that is not relevant?
3. Identify the literary mechanisms used. Are they the most effective?
4. Determine the persuasiveness of the report. How could the report be more persuasive, if possible?
5. If you were proofreading this report, would you request any additions, deletions, or changes?

CHAPTER 3

Principles of Clear Technical Writing

Writing clearly and concisely gets the job done.

Technical writing should effectively communicate an idea, concept, or information. This text teaches you the fundamentals of communication with a purpose and of writing clearly and concisely. The principles of grammar and sentence structure are discussed in this text only to the extent that they have an impact on effective technical writing.

This chapter discusses the principles of using a natural style to make your writing more understandable, communicating effectively by being direct, writing convincingly by using proper expression, writing effective instructions, and using the proper verb tense for your reports.

When you write, be concerned with the background and perspective of your readers and how they might inappropriately interpret your message. When you have any doubt concerning the effectiveness of your message, revise its text to eliminate this doubt.

KEEPING A NATURAL WRITING STYLE

A natural writing style will increase the readability of your reports and eliminate the need for repeat readings to understand your message. Follow the guidelines in this section (see Figure 3–1) to increase the readability of your writing:

Structure

1. Use lists with headings and introductory remarks. Identify the purpose of the list using headings and introductory remarks. Numbers or letters in sequence (e.g., 1, 2, etc., or a, b, etc.) precede the items in the list when the

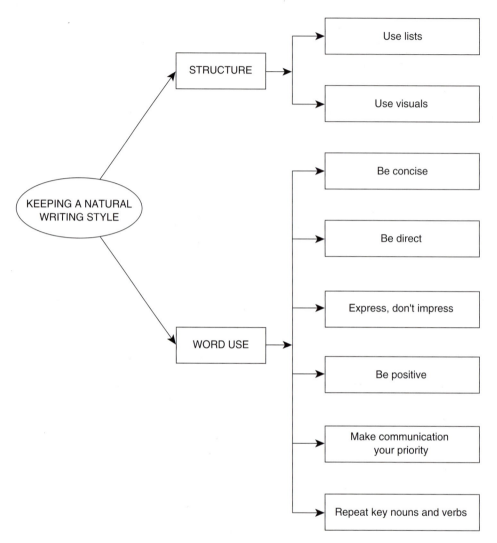

FIGURE 3–1
Keeping a Natural Writing Style

sequence of operation or chronology of events is important. Numbers or letters also provide easy identification. Otherwise, use bullets or other graphics. Parallel grammatical structure is used for all items in a list; for example, all items may be imperative statements, statements of facts, or quantities of articles. Also, all items should use the same tense (i.e., past, present, or future).

Examples

Not parallel: Use a fine file to create a smooth surface. [Imperative]
Painting the surface with matte paint will create a dull finish. [Future tense]
Shellacking the surface prolongs the life. [Present tense]

Parallel: [All imperative]
Use a fine file to create a smooth surface.
Paint the surface with matte paint to create a dull finish.
Shellac the surface to prolong the life.

Parallel: [All present tense]
Using a fine file creates a smooth surface.
Painting the surface with a matte paint creates a dull finish.
Shellacking the surface prolongs the life.

2. Use visuals.   Use graphs, charts, tables, and photographs to simplify explanations and descriptions and to create mental images.

Word Use

1. Be concise in your choice of words.   Carefully select the words or phrase to express your idea in as few words as possible.

Examples

Instead of writing *a large number of*, write *many.*
Instead of writing *in the course of*, write *during.*

2. State your ideas as directly as possible.   Avoid circumventing the main points or adding unnecessary explanations to prevent confusing the reader.

Example

Indirect: Due to uncertainties in the weather, it is difficult to predict when the first flight will be. However, the preference is for tomorrow.

Direct: If the weather permits, the first flight will be tomorrow.

3. Write to express, not to impress.   Pompous language can easily camouflage your intended meaning. Industry jargon and acronyms should not be used unless you are sure that your readers will understand their meanings.

Examples

Pompous Language
Instead of writing *contemplate*, write *consider* (e.g., "We will *consider* revising the design").
Instead of writing *endeavor*, write *attempt* or *try* (e.g., "The technician *attempted* to increase the test frequency").

Jargon
Instead of writing *facilitator*, write *administrative assistant*.
Instead of writing *RIF*, or *reduction in force*, write *layoff*.

4. Be positive in your information.   Use of the word *no* or *not* may require judgment or interpretation of the meaning by readers. Rather, positively state what you mean.

Example

Negative: This acid is *not effective* for chemical etching.

Positive: This acid is *too weak* for chemical etching.

5. Make communication of your ideas your priority.   Rules of grammar do not need to be rigidly followed when violation of these rules simplifies the understanding of your ideas. For example, sentences may end with a preposition, paragraphs may include only one sentence.

Example

Acceptable: Which supplier should the parts be shipped *to?*

6. Repeat key nouns and verbs whenever necessary.   Changing nomenclature for the purpose of creating interest is potentially confusing to readers.

Example

Inconsistent: The surface of the *movable work platform* was usable, even though the wood needed to be repaired. . . . Therefore, salvaging the *scaffold* was feasible.

Clear: The work surface of the *scaffold* was usable, even though the wood needed to be repaired. . . . Therefore, salvaging the *scaffold* was feasible.

COMMUNICATING EFFECTIVELY

The following guidelines (see Figure 3–2) will help you to communicate effectively.

Contents

1. Use building block organization.   Present your headings and information in a sequence that builds on the information previously presented and facilitates selective reading by readers not interested in reading the entire report. For example, in an analytical report, the sequence of headings may be *Introduction, Given Data, Assumptions, Calculations, Results*, and *Conclusions*. This building

FIGURE 3–2
Communicating Effectively

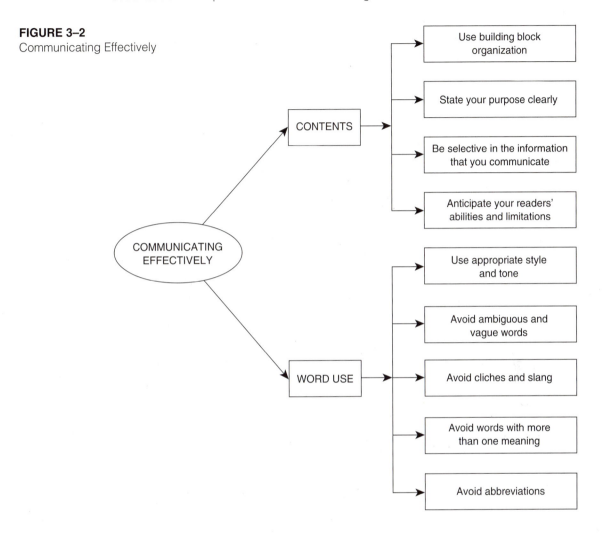

block sequence for presentation of information begins with a statement of the purpose of the report, continues with the analysis, and concludes with its findings.

2. State your purpose clearly.   Make sure readers know what you hope to accomplish at the beginning of the document.

Example

The purpose of this report is to demonstrate compliance of the HEPA 12 air filter with Section 16.04 of the MIL-STD 12, Rev. C dated 12 Dec 1988.

3. Be selective in the information that you communicate.   Tell the readers everything they need to know. However, be careful to include only relevant information that helps them to understand your message.

4. Anticipate the abilities and limitations of your readers. Include sufficient details and explanations for your readers to understand those items for which they have little knowledge or background. Omit details and explanations that may undermine your readers' abilities. When in doubt, include rather than omit details and explanations.

Word Use

1. Use appropriate style and tone. Requests should be friendly, instructions should be in the imperative, and conclusions should be affirmative. Demanding, vindictive, and derogatory tones cause the reader to become defensive and reluctant to objectively read and understand your message. Therefore, these tones are inappropriate for technical writing.

2. Avoid ambiguous and vague words. They create uncertainty and confusion for the reader.

Examples

Ambiguous: The flow of lava was *affected*. (Was the flow increased or decreased?)

Clear: The flow of lava was *decreased*.

Vague: *It* created a glossy appearance.

Clear: *The lacquer* created a glossy appearance.

3. Avoid clichés and slang. Clichés and slang give readers the impression that the writer is incapable of clear expression.

Examples

Avoid clichés such as *in any event, it goes without saying*, and *last, but not least*.

Avoid slang: Instead of writing *down the road*, write *later*.

Instead of writing *cut corners*, write *reduce*

Instead of writing *a drop in the bucket*, write *an insignificant amount*.

4. Avoid words with more than one meaning. They may be misinterpreted or unclear and require the reader to read further to understand their meaning.

Examples

Misinterpreted: *Since* the component was rejected, a new manufacturing process was developed. (Was the process developed as a result of the rejection of the component, or after the component was rejected?)

Clear:	*Because* the component was rejected . . . . (Demonstrates cause and effect.) *After* the component was rejected . . . . (Demonstrates chronology.)
Misinterpreted:	This is the *last* carburetor to be installed. (Was this the most recent or final carburetor to be installed?)
Clear:	This is the *most recent* carburetor to be installed. (Demonstrates an ongoing process.) This is the *final* carburetor to be installed. (Demonstrates termination of the process.)
Unclear:	*Once* we increased the strength, we had no additional failures. (Does *once* mean "only one time" or "after"?)

5. Avoid abbreviations unless the word is commonly abbreviated. Abbreviations can be misinterpreted by the reader.

Examples

Instead of writing *elect. engr.*, write *electrical engineering* or *electrical engineer* (whichever you mean).

Instead of writing *v.p.*, write *vice president*.

However, *etc.* (for *et cetera*) is an acceptable abbreviation, but often is used in parentheses only. Outside parentheses, *etc.* is often replaced with a term such as "and so on."

WRITING CONVINCINGLY

Effective technical expression begins with the formation of coherent sentences and appropriate syntax (the combination of words and phrases to form sentences in a direct and natural manner). These elements are essential to communicate ideas and to persuade readers. Also see Figure 3–3.

Sentences

1. Begin sentences with the central idea or concept. This focuses the reader's attention on this idea or concept. Any necessary explanations follow the central idea or concept.

Examples

Misleading beginning:	The increased density of the air means that the efficiency of the engine is increased.
Appropriate:	The efficiency of the engine increases with the increased density of the air.

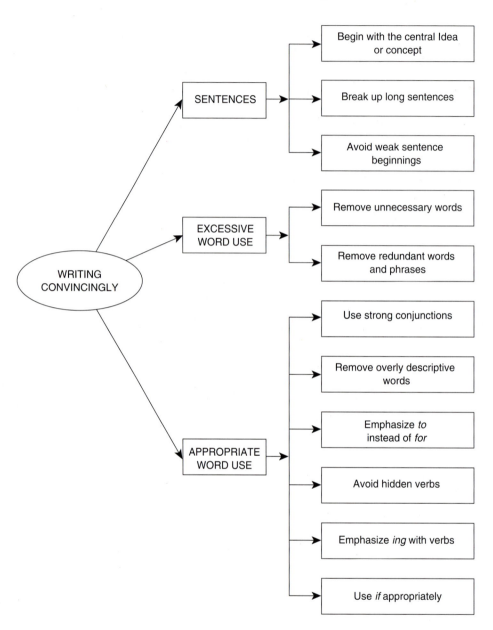

FIGURE 3–3
Writing Convincingly

Misleading beginning:	When the parts were strengthened, the number of reported failures decreased significantly.
Appropriate:	The number of reported failures decreased significantly when the parts were strengthened.

2. Break up long sentences.   Concepts with more than one central thought are effectively communicated with shorter sentences.

Examples

Long:	A seminar concerning the use of our new Microstation computer program will be presented in the conference room next Tuesday at 2 p.m. for all engineers and drafters.
Appropriate:	A seminar concerning the use of our new Microstation computer program will be presented in the conference room next Tuesday at 2 p.m. All engineers and drafters must attend.
Long:	It is our design philosophy that axial members of all structural systems have redundant load paths to minimize the possibility of structural failure and that the ends of these members have rigid joints to reduce deflection.
Appropriate:	It is our design philosophy that axial members of all structural systems have redundant load paths to minimize the possibility of structural failure. Also, the ends of these members should have rigid joints to reduce deflection.

3. Avoid weak sentence beginnings.   Unnecessary beginning phrases detract the reader's attention.

Examples

Weak:	*It was the* resilience of the material that prevented it from shattering.
Appropriate:	*The* resilience of the material prevented it from shattering.
Weak:	*As far as the rivets are concerned, they* comply with the appropriate codes.
Appropriate:	*The rivets* comply with the appropriate codes.

Excessive Word Use

1. Remove unnecessary words.   Remove words and phrases that do not clarify ideas or concepts.

Examples

Unnecessary:	We purchased the computer *for the purpose of* increasing the efficiency of the department.

Appropriate: We purchased the computer *to* increase the department efficiency.

Unnecessary: The engine became overheated *because of the reason that* the
fanbelt was not properly installed.

Appropriate: The engine overheated *because* the fanbelt was not properly
installed.

2. Remove redundant words and phrases.   When an idea is stated redun-
dantly, it detracts from the remainder of the sentence.

Example

Redundant: The unlit, dark room was hazardous.
Appropriate: The unlit room was hazardous.

Appropriate Word Use

1. Use strong conjunctions.   Conjunctions such as *however, but,* and
because reveal interrelationships between connected thoughts to the readers.
Weak conjunctions such as *and* conceal these relationships. Semicolons can be
used with these conjunctions to separate interrelated thoughts that otherwise can
be expressed independently.

Examples

Weak: The surface of the housing was corroded, *and* the steel was exposed.

Strong: The surface of the housing was corroded *because* the steel was
exposed.

Weak: The deck of the bridge was resurfaced regularly, *and* the approach
ramp was poorly maintained.

Strong: The deck of the bridge was resurfaced regularly; *however,* the
approach ramp was poorly maintained.

2. Remove overly descriptive words.   Words that are not simply understood and require interpretation detract from the intended meaning of ideas.

Examples

Overly descriptive: Each computer station is *equipped* with storage space.

Appropriate: Each computer station *includes* storage space.

Overly descriptive: To prevent corrosion, all equipment should be *enclosed*
when not in use.

Appropriate: To prevent corrosion, all equipment should be *covered*
when not in use.

3. Emphasize *to* instead of *for.* *To* followed by a verb makes your statement active.

Examples

Weak: The metallurgists have 2 weeks *for completion* of the research.

Strong: The metallurgists have 2 weeks *to complete* the research.

Also, in an imperative sentence that emphasizes cause and effect, placing the desired result (i.e., effect) in front of the cause emphasizes this desired result.

Weak: *For detection* of surface cracks, use dye penetrant.

Strong: *To detect* surface cracks, use dye penetrant.

4. Avoid hidden verbs.   Expressing the activity as an action verb emphasizes the activity rather than the actor of this activity.

Examples

Weak: The chromium plating *provides protection* against corrosion.

Strong: The chromium plating *protects* against corrosion.

Weak: This procedure *is applicable* to all departments.

Strong: This procedure *applies* to all departments.

5. Emphasize *ing* with verbs.   Expressing the action after an article (e.g., *a, the*) detracts from the importance of this action.

Examples

Weak: *The removal* of the damper increased the flow of air.

Strong: *Removing* the damper increased the flow of air.

Weak: *The implementation* of the new design procedure will require the cooperation of all supervisors.

Strong: *Implementing* the new design procedure will require the cooperation of all supervisors.

6. Use *if* appropriately.   Use *if* only for those actions that are uncertain to occur, not those actions whose time to occur is uncertain.

Examples

Inappropriate: *If* the ambient temperature is below 60°F, the rubber compound does not cure properly. (This action is certain to occur, only the time to occur is uncertain.)

Appropriate: *When* the ambient temperature is below 60°F, the rubber compound does not cure properly.

Appropriate: *If* the test is not completed tomorrow, we will work on Saturday. (It is uncertain that the test will be completed tomorrow.)

WRITING INSTRUCTIONS*

Write instructions as commands in the sequential order of performance.

- Each step of an operation or procedure usually begins with an imperative verb such as *measure, place,* or *cut.* Number the steps in the order of the operation. Specify any information necessary to perform a step or series of steps (e.g., a warning, or a time limitation) before that step or series of steps. Do not include statements that are informational only unless the information is required to understand the instruction that follows.

Examples

Informational only: *You may adjust* the speed of the ram, *if necessary.*

Imperative: *If necessary, adjust* the speed of the ram.

Dangerous: The red lever shall be held in its lowest position while performing Step 6.2 *to prevent an explosion.*

Safe: *WARNING: To prevent an explosion,* hold the red lever in its lowest position while performing the following step (Step 6.2).

TENSE

Because the subject of your report is current, use the present tense to address any event that relates to the project. Also address other parts of the report in the present tense.

Examples

Incorrect tense: The environmental load test of the silicon computer chip *simulated* a 10-year life.

Appropriate: The environmental load test of the silicon computer chip *simulates* a 10-year life.

In the Introduction section of the report:

Incorrect tense: This report *will evaluate....*

---

*Discussed in greater detail in Chapter 5.

Appropriate: This report *evaluates....*

In the Discussion of Results section of the report:

Incorrect tense: The analysis *confirmed....*

Appropriate: The analysis *confirms....*

Use the past tense for an event that was completed before the conception of the project and the report.

Example

Appropriate: This modification of the steel strut is to prevent future failures similar to the one that *occurred* last March.

Use the future tense for future or hypothetical events that are the topics of proposals and environmental impact reports. Using *would* and *should* for predictable outcomes is cumbersome. Rather, switch to the present tense.

Example

Cumbersome: If the construction of this parking structure is approved, traffic on Main Street *would increase* by approximately 10 percent. An increase in traffic *would increase* road maintenance expenses.

Appropriate: If the construction of this parking structure is approved, traffic on Main Street *would increase* by approximately 10 percent. An increase in traffic *increases* road maintenance expenses.

Table 3–1 summarizes all of the points discussed in this chapter. After writing your technical report, you can ensure your report's clarity by reviewing its conformance with these points.

KEY CONCEPTS

- Incorporating lists and visuals into your text, using a natural writing style, increases the readability of your writing.
- Adopting the perspective, knowledge, and background of your readers will help to determine the details and explanations that should be included in your report.
- Selecting words and phrases that are not vague, redundant, or have more than one meaning contributes to effectively communicating your message.

TABLE 3–1
Checklist for Reviewing the Clarity of Your Technical Reports

Keeping a Natural Writing Style

Structure
1. Are lists with headings used? Do these lists include introductory remarks?
2. Are visuals included when they would be helpful to the reader?

Word Use
1. Are your words concise?
2. Are your ideas stated directly?
3. Have you avoided pompous language?
4. Is your information stated positively rather than negatively?
5. Is communication of ideas your priority?
6. Are key nouns and verbs repeated when necessary?

Communicating Effectively

Contents
1. Did you build headings and information on ideas and concepts previously presented?
2. Did you state your purpose clearly?
3. Did you include only relevant information?
4. Were the abilities and limitations of your readers considered?

Word Use
1. Is the style and tone appropriate for this type of report?
2. Are ambiguous and vague words avoided?
3. Are cliches and slang avoided?
4. Are words with only one meaning included?
5. Are abbreviations avoided unless commonly abbreviated?

Writing Convincingly

Sentences
1. Do sentences begin with the central idea or concept?
2. Are sentences short enough for readers to understand easily?
3. Do sentences have strong beginnings?

Excessive Word Use
1. Have unnecessary words been eliminated?
2. Have redundant words and phrases been eliminated?

Appropriate Word Use
1. Are strong conjunctions used to interralate thoughts?
2. Are overly descriptive words avoided?
3. Is *to* followed by a verb, emphasized rather than *for* followed by a noun ending in *ing?*
4. Are hidden verbs avoided?
5. Are actions emphasized with *ing* rather than introduced with an article followed by a noun ending in *al* or *tion?*
6. Is *if* used only for those actions that are uncertain to occur?

Writing Instructions
Do all instructions begin with an imperative verb?

Tense
Is the present tense used for all items discussed in the report, the past tense for all actions completed before the conception of the report, and the future tense for all future or hypothetical events?

STUDENT ASSIGNMENT

1. The following words can effectively communicate certain ideas, concepts, and information. However, they are sometimes used to impress when a simpler word is more appropriate. Substitute a simpler word for each of the following:

depict	utilize
facilitate	endure
magnitude	transmit
retain	terminate
equitable	impeccable

2. Revise the following sentences using the principles learned in this chapter to effectively communicate the writer's intent.
   a. The group will decide the question of whether more office space will be needed.
   b. Please indicate in your letter of transmittal the number of steel billets that you have a preference for.
   c. The foreman together with his crew were responsible for start-up of the plant.
   d. The flashing green light is indicative of the readiness of the vehicle to function.
   e. Access ramps are geared for the handicapped.
   f. It was only one tooth on the gear that did not comply with the tolerance requirements.
   g. The steel bolts with the nuts have been properly plated.
   h. The negligible bending stress in the member means that the required size of the member can be reduced.
   i. The research group began the testing procedure.
   j. The technicians have 5 weeks remaining for qualification of the components.
   k. The handbook will be purchased by every engineer.
   l. There are two groups that are responsible for the operation of the reentry vehicle.
   m. The circuit board is intended to be built before the end of the year, while the project is not required to be completed for several years.
   n. The videotape was about one hour long.
   o. The size of the proposed plant is up in the air.
   p. A meeting was held with our client for discussion of critical speeds.
   q. The structural analysis has validity for the design.
   r. The framers worked overtime after hours and did a first-class job.
   s. The aircraft landing gear assembly tires need to be replaced immediately.
   t. The client's request was complied with by our designers.
   u. The bottom line is that the design of the existing facility will be a model for designing the new facility.

    v. Use a micrometer for inspection of critical components.

    w. Concerning the crane, it has sufficient reach to complete the job.

    x. The catalyst helps speed up the chemical reaction.

    y. To ensure proper operation, the tolerances of the components were reduced across the board.

    z. This filter is not effective if the particle size exceeds 5 microns.

3. Revise the following instructions:

    a. The load developed by the ram shall be increased in 5,000-lb increments to 50,000 lb.

    b. Place the soil sample in the oven to dry after compacting it in accordance with the procedure on page 57 of this manual.

    c. Press the ON button. Make sure that your hands are away from the anvil before performing this operation.

CHAPTER 4

Rules of Practice for Technical Writing

Proper use of accepted standards of practice emphasizes your message.

The rules of practice for technical writing allow writers to emphasize important data, phrases, and concepts in a manner that all readers understand; therefore, they are tools that writers use to achieve clarity. Because many of the rules of practice are inconsistent with each other, writers must consider the perspective, background, and knowledge of the readers in selecting the most effective rules to use. However, when these rules have been selected, they should be used consistently throughout the report.

This chapter provides general guidelines for use in writing reports. In your professional life, you will probably also consult the style manual produced for your specific discipline as well as any writing guidelines specified by your employer. You may also wish to consult a general style manual, such as *The Chicago Manual of Style* (14th Ed., Chicago, University of Chicago Press, 1993). This widely used style manual includes specific sections referring to scientific and technical writing.

ABBREVIATIONS AND ACRONYMS

The primary purpose of using abbreviations and acronyms (a word formed from the initial letters of a name, e.g., Department of Energy expressed as DOE) in technical writing is to simplify the reading, not the writing, of the text material. For example, if you referred to the lengthy title Nuclear Regulatory Commission only once or twice, it makes the most sense to write out the full name even though it is reasonable that most readers are familiar with the shortened form, NRC. However, when you intend to make the reference frequently, it should appear the first time as Nuclear Regulatory Commission (NRC), which alerts the reader that the shortened form, NRC, will appear later in the text.

On the contrary, when you are certain that all readers understand the meaning of an abbreviation or acronym, this abbreviation or acronym can be introduced into your writing in the shortened form without being defined. For example, the writer of an internal memo within the National Aeronautics and Space Administration may refer to this administration as NASA.

Use the following additional principles for abbreviations and acronyms:

General

- Include an abbreviations and acronyms section at the beginning of the report when you expect to use many different shortened forms. It should list all shortened forms in alphabetical order followed by the full words or names. The shortened form only can then appear in the text without further definition.

Example

ASTM   American Society for Testing and Materials
ERDA   Energy Research and Development Administration

- Begin an abbreviation with a capital letter only when the full word begins with a capital letter (e.g., K for degrees Kelvin but *vert.* for *vertical*).

- Do not begin a sentence with an abbreviation.

- When ending a sentence with an abbreviation, only one period is used. However, this may mislead the reader directly into the following sentence without a break in thought. Therefore, to avoid confusion, it is good practice to avoid abbreviations at the end of a sentence and either use the full word or revise the sentence to place the abbreviated word elsewhere.

Symbols

- Use the symbols # for pound(s) and % for percent in numerical analysis, tables, and figures. Also, for emphasis, the symbol % may be used in text when preceded by digits.

  Do not write the symbol # to abbreviate *number* because, in technical reports, # is interpreted as *pounds*.

- Use the symbols $ and ¢ for dollars and cents when the amount is expressed in digits rather than words.

Examples

Instead of writing 98 cents, write 98¢.

Instead of writing 16 dollars and 25 cents, write $16.25.

Units of Measurement

- When standard units of measurement are abbreviated, write them in the singular, and without a period. However, always use a period when abbreviating *inches* because it can be misread as the word *in* rather than *inches*, as intended (see Figure 4–1).

Examples

Wrong: 120 lbs of force

Correct: 120 lb of force

Wrong: 12.0 in long

Correct: 12.0 in. long

However, when units of measurement are not abbreviated, they are expressed in the plural when the number is greater than 1 (e.g., write 0.88 inch, but 1.02 inches).

Also, singular or plural units are determined by the number, rather than by the value of the number, specified before the units. For example, *0.75 kilogram* (which represents 750 grams) is singular because less than 1 kilogram is specified; and *50 centimeters* (which represents 0.50 meter) is plural because more than 1 centimeter is specified.

- Abbreviate standard units of measurement only when they are preceded by Arabic numbers (e.g., write *35 gal*, and *1800 rpm*).

- It is not necessary to define standard units of measurement when they are commonly used and understood (e.g., write *cu yd*).

- Use symbols for minutes (′) and seconds (″) of angle measurements only. Do not use these symbols to represent minutes and seconds of time or feet and inches of length. However, you can use ° as the symbol for degrees of angle measure or degrees of temperature (see Figure 4–2).

FIGURE 4–1
Abbreviation of Units of Measurement

- Singular and without a period except *in.* when abbreviating inches
- Only when preceded by Arabic numbers

FIGURE 4–2
Measurement Symbols

Angles:	°, ′, ″
Time:	h, m, s or hr, min, sec
Length:	ft, in.
Temperature:	°

Examples

Angle:	171° 41′ 30″
Time:	15 m 45 s [or] 15 min 45 sec
Length:	10 ft 9 in.
Temperature:	72°

NUMBERS

Numbers can be expressed in technical reports as words (e.g., twenty-three) or digits (e.g., 23). Many of the common practices are inconsistent with each other. The following conventions (see Figure 4–3) are those typically used by most technical writers:

General

- Select a convention to use in the report. To avoid confusing the reader, you must be consistent in its use.
- Use digits for the following:

  Numbers of measurement or data (e.g., *2.5 ft*. Counted numbers can be expressed either way: *twenty-three beams* or *23 beams*).

  Any number 10 or greater (e.g., *14 boilers*).

  Any number less than 10 when units of measurement are included (e.g., 8 ft/sec).

  Any number less than 1.0. A zero is placed in front of the decimal point (e.g., *0.375*) unless it is not possible for the number to equal or exceed 1, as in probabilities (e.g., *p* < *.05*).

  For greater impact on the reader when the quantity is significant (e.g., *6 failures*). Most readers understand and remember digits more easily than words. This concept is especially useful in business correspondence.

  Money (e.g., *$6* or *$69.95*). However, money is expressed in words for approximate amounts (e.g., *fifteen dollars*).

FIGURE 4–3
Number Expression Guidelines

Digits	Words
• Any measurement or data	• Any approximation
• Any number 10 or greater	• Any number less than 10 without units
• Any number less than 10 with units	• The beginning of a sentence
• Any number less than 1.0	
• When the numerical value of a quantity is important	
• Money, when exact	

- Use words for the following:

Any number that is an approximation (e.g., less than twenty-five hundred miles).

Any number less than 10 when units of measurement are not included.

The beginning of a sentence when the sentence begins with a number.

Example

Wrong: 206 miles north of the city. . . .

Correct: Two hundred and six miles north of the city. . . .

- In a series, use the same convention for all numbers, either digits or words. The convention is usually determined by the longest number in the series. (e.g., use digits to write *1 screwdriver, 1 wrench, 416 screws, 416 nuts, and 12 sheets of steel*, because 416 is greater than 10.)
- When two numbers are expressed together, express one in digits and the other in words to avoid confusing the reader. Usually, the shorter one is expressed in words. (e.g., *28 six-penny nails* and *twelve 3/16-inch screws.*)
- You may use engineering notation (mixed digits and words) for very large and very small numbers. (e.g., *38 thousand screws* and *12 microamps.*)
- For contracts and legal documents, use words followed by digits in parenthesis. (e.g., *Four hundred and twenty-five (425) cans of paint.*)
- Metric units

It is sometimes necessary in a report to include units of measurement in the metric (SI [International System of Units]) system, as well as the English system. The convention in the United States is to include the metric equivalent in parentheses after the English units (e.g., 12.8 ft (3.90 m)). For other industrialized countries where the metric system is used, include the English equivalent in parentheses after the metric units (e.g., 3.90 m (12.8 ft)).

When calculating the metric equivalent of English units, use the same number of significant digits in the metric equivalent as in the English units.

Example

Wrong: 346 gal (1.3096 m^3)

Correct: 346 gal (1.31 m^3) [i.e., both English and metric include only three significant digits.]

EQUATIONS AND CALCULATIONS

Equations and calculations (herein referred to as equations) are frequently included in research reports and engineering and design analyses. Use the following guidelines to present them in formal reports as an integral component of the text material:

- Using words, introduce an equation in the text before presenting the mathematical symbols and numerical calculations.
- Either indent the beginning of an equation from the left margin or center the equation on the line.
- Introduce an equation using mathematical symbols (i.e., variables, coefficients, and exponents) before numerically solving it.
- When showing the numerical solution to an equation, show the numerical calculations in equation form only and the final answer. Do not include intermediate arithmetic steps.
- When an equation is referenced in the text, place an equation number or lowercase letter in parentheses inside the right-hand margin. The reference in the text to the equation number always appears subsequent to the appearance of the equation number.
- Leave one space before and after all operation signs (i.e., $+$, $-$, $\times$, $\div$, and $=$). However, do not leave a space between the minus sign indicating a negative quantity and the quantity (e.g., write -8 *ft/sec*, not -8 *ft/sec*).
- When several lines of equations are included, line up the equal signs. Therefore, the left margins may be irregular.
- When defining mathematical symbols used in an equation, include the units of measurement in parentheses after each definition. Use a list for defining more than one symbol.

Example

The combined compressive stress is defined by

$$f_c = \Sigma \, P/A + Mc/I \tag{6}$$

where:

f_c = combined compressive stress (lb/in.^{2})

ΣP = sum of the axial loads (lb)

A = area of the cross section (in.^{2})

M = bending moment about the x-axis (in.-lb)

c = distance to the compression fiber (in.)

I = moment of inertia about the x-axis (in.^{4})

Therefore,

$$f_c = 82{,}000/6.19 + 83{,}200(4.95)/106.3, \text{ and}$$

$$f_c = 17{,}120 \text{ \#/in.}^2$$

- When an equation requires more than one line

1. Separate the equation only at a plus ($+$) or a minus ($-$) sign. Do not separate terms that are multiplied, divided, or otherwise operated on. However, an expression within parentheses can be separated at a plus or minus within these parentheses.

2. Begin the second and subsequent lines with the plus or minus sign aligned with the first symbol or number of that side of the equation directly above it.

Example

Wrong: $$F = 32.2(65.5 + 4.87)^2 + (45.0 + 21.9 - 6.0)[0.00452 + 2.00(52.9 + 79.9)^{1.33}]^{\#}$$

Correct: $$F = 32.2(65.5 + 4.87)^2 + (45.0 + 21.9 - 6.0)[0.00452 + 2.00(52.9 + 79.9)^{1.33}]^{\#}$$

COMPOUND TERMS

Compound terms, or hyphenated words, help a writer clarify meaning and give emphasis that would otherwise be difficult to achieve.

Hyphenated words help a writer to clarify meaning by identifying

- Compound adjectives modifying a noun: *double-acting pistons* and *two-way street*
- Compound nouns: *y-coordinate* and *passenger-miles*
- Compound verbs:

  The mechanic reverse-flushed the radiator.

  The technician heat-treated the steel.
- Closely related words in a phrase: *tool-and-die* and *black-and-white photograph*

Hyphenated words also help to avoid misinterpretation of phrases such as *heavy metal pollution*, which can be intended to be *heavy-metal pollution* (pollution by heavy metals), or *heavy metal-pollution* (heavy pollution by metals).

SPECIAL APPLICATIONS FOR COMMAS AND SEMICOLONS

Because technical writing includes content that is different from literary and journalistic writing, commas and semicolons are used for these special applications (see Figure 4–4):

FIGURE 4–4
Special Applications for Commas and Semicolons

Commas	Semicolons
• Before the last element in a series	• To separate groups in a series
• In numbers greater than four digits	• To separate numbers in a series when any number includes a comma

- When only two elements are used in a series, a comma is not inserted between these two elements. However, when more than two elements are used in a series, a comma is usually inserted between the final two elements before the word *and* or *or* for clarity. This convention is particularly helpful in reading technical writing because elements that appear in a series may include compound terms (e.g., *cause and effect or nuts and bolts*). This comma between the last two elements in a series that consists of more than two elements alerts the reader that the series is ending. This convention is used in technical literature published by the U.S. government and by most technical writers.

Examples

Wrong: Each component is fabricated from either *steel, or* aluminum.

Correct: Each component is fabricated from either *steel or* aluminum.

Vague: Each participant must have one calculator, *pencil and paper and* one straight-edge.

Better: Each participant must have one calculator, *pencil and paper, and* one straight-edge.

- When elements in a series contain internal punctuation, a semicolon is used to separate the elements.

Example

Confusing: Well-written technical reports usually include the following: correct punctuation, grammar, and syntax, mathematical accuracy of calculations, reasonable assumptions, logical conclusions, and feasible recommendations.

Improved: Well-written technical reports usually include the following: correct punctuation, grammar, and syntax; mathematical accuracy of calculations; reasonable assumptions; logical conclusions; and feasible recommendations.

- A comma is used to separate digits for numbers greater than four digits only. A comma is used in a four-digit number only when that four-digit number is included in a column or in a sentence that includes other numbers greater than four digits. Also, in a column of numbers, all decimal points line up with each other.

Examples

Instead of writing 4,647, write 4647 in text.

But, write 73,987.

Also, write

23,863.28

6,028.61

671.43

37,829.50

68,392.82

- When any number included in a series of numbers includes a comma, the numbers in that series are separated by semicolons.

Examples

76,287; 4876; 3650; 539, and 49,902

or

356, 82, 749, 271, and 86

- All other uses of commas and semicolons in technical writing are the same as for literary and journalistic writing.

KEY CONCEPTS

- The rules of practice for technical writing are your tools for emphasizing and clarifying the important technical concepts of your report.
- The rules of practice for technical writing promote clarity and allow you to emphasize your message easily.
- The perspective, knowledge, and background of the readers are considered in selecting the most appropriate rules and conventions to use in the report.

STUDENT ASSIGNMENT

Revise, if necessary, the following sentences to reflect the technical concepts related to numerical concepts, units of measurement, and equations:

1. The lifting is performed by four one-hundred ton cranes.

2. Several gal of gas are required for the trip.

3. The gap between the beams was 0.50 in..

4. The conveyor belt is 142 ft (43.28 m) long.

5. The requisition asks for twenty-seven machinists to be hired for the project.

6. The assembly is held together with 16 five-eighths-in. bolts.

7. We used 4 wire-mesh screens and two doorframes.

8. The bending moment at the center of the beam is 520 ft-lbs.

9. 50 microamps are required for the relay.

10. Our profit margin for the year was 18 percent.

11. The prototype of the Model MJ-2 helicopter achieved a maximum speed of 160 MPH.

12. The shop has 42 used lathes for sale.

13. It costs almost $3000 per week to run the operation.

14. The project is approximately 25% complete.

15. The Air Force will purchase four prototype airplanes for testing.

16. The eight cylinder engine was trouble-free.

17. Enclosed is your order for four hundred digital thermometers.

18. The half spent fuel cell can be dangerous.

19. Use a #8 sheetmetal screw for Step 7.

20. We completed 16 thousand hours without an accident.

21. Thirty-two in/sec is too fast to maintain control.

22. The instructions were in an easy to-read format.

23. The natural frequency of the system is 38 hz.

24. We purchased 100 pounds of sulfuric acid.

25. The slab is .75 in. thick.

26. Two 40-pound bags were adequate for the job.

27. $M_2 = aP(1 - 2/\pi)$ (Eq. c)

28. The height of the wall is 96 in from the floor to the ceiling.

29. The motor speed is 3500 revolutions per minute (rpms).

30. The list of equipment includes two tractors, one backhoe, one grader and, 28 hand shovels.

31. The dry weight of the vehicle is 8240#.

32. The size of the housing, 12′ 6″ × 9′, is too large to fit in the enclosure.

33. The warm environment design temperature is 150° Fahrenheit.

34. 6,400 gal of fuel were consumed.

Elements of the Technical Report

Section II presents typical components of technical reports: the cover materials, the report sections discussed under report headings, the visuals, and desktop publishing.

Each report section discussed includes simulated material and the structure and content of writing typical for that section. These sections use different style formats to show the different visual effects on readers. Section numbers, when included with the headings of these sample sections, are intended to be in context with a report not included in this text; therefore, the section numbers of the sample sections have no relevance to this text, or other sample sections in this text.

A report, when completed, includes many of the simulated report sections shown in the chapters in this Section II. Although these report sections are discussed independently, they are integrated into one to compose a coherent report.

The order of presentation of Chapters 5 through 8 represents the order in which these sections and materials are usually prepared by writers. However, the order of presentation of sections and materials within each chapter represents the order they are usually presented in these writers' reports. Section III discusses the order in which the appropriate sections are included in reports.

The simulated sections of professional reports in Section II include critiques on the *structure* and *content* of the writing.

Structure includes

Format and organization

Use of visuals

Parallelism of expression

Consistency and conformance to standard practices

Tense

Appropriate technical level for the audience

Whereas, content includes

Understanding of purpose

Addressing the needs of the audience

Addressing the appropriate issues

Because of their importance, persuasiveness, clarity and conciseness of expression, syntax, and tone are also included in these critiques.

The exercises at the end of each chapter include samples of sections for you to rewrite and sections of reports for you to write to give you experience.

SELECTION OF REPORT HEADINGS AND CONTENT OF SECTIONS

Technical reports convey information, state conclusions, and make recommendations arising from studies. However, readers need to be carefully guided from the initial facts or data that are the bases for these studies, to the presentation of the information, final conclusions, and recommendations.

Reports that are only one to two pages long can be written in typical paragraph format, for example, introduction, body, conclusion. However, longer reports require sections with headings to direct the readers' attention and help them locate material.

A preliminary organization of sections with headings forms the framework for writing. The contents or functions of the sections with headings are revised, if necessary, as the reports are written. Sometimes, sections need to be added or eliminated when the information is adequately presented elsewhere in the report.

Frequently, reports have more than one reader. Headings guide readers through the report and help them selectively read only pertinent sections. Also, because readers often have different perspectives from each other, and direct their reading to pertinent sections only, information presented in each section of the report must be adequate for readers to understand.

The structure and content of report sections are discussed extensively in Section II of this text. Writers select applicable report sections for their reports and adapt these sections to fulfill the needs of their audiences.

CHAPTER 5

The Body of the Report

The body of the report demonstrates the final conclusions and recommendations.

The body of the report validates the conclusions and recommendations of the report. The information presented in the body must be adequate to predict the outcome or ending of the report. Usually, only technical readers read this section of the report. However, this is the most important section for the client (the purchaser of a product or service) reviewing the report for approval and acceptance. Also, when the report determines a course of action, and the result of the action differs from the prediction in the report, the body of the report will be carefully reviewed for errors in methodology, theory, design parameters, technical requirements, assumptions, and calculations.

Information, data, and calculations in the body of the report must be fully, accurately, and sequentially presented so that it can be understood by any professional competent in an appropriate discipline, because this material is subject to critical review and can become the potential source of a lawsuit.

The body of the report may include many of the following sections (see Figure 5–1), shown in the order in which they typically appear:

- Discussion of methodology
- Theory
- Design parameters
- Technical requirements
- Assumptions
- Technical description
- Instructions and procedures
- Evaluation and analysis
- Acceptance criteria

FIGURE 5–1
Sections of the Body of the
Report

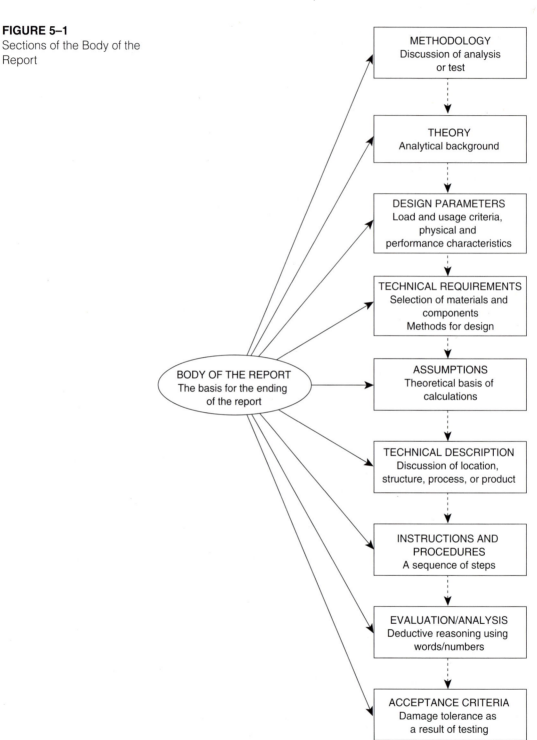

The body of the report is usually written first because its outcome determines the text of the other parts of the report.

DISCUSSION OF METHODOLOGY

When the report has an analytical or test component, and the methodology is unusual, the methodology is explained. This section discusses the philosophy of the methodology; it does not, however, include the design requirements, theory, or details reserved for other sections of the report.

In the following discussion of methodology, only the philosophy of the analysis and design is discussed (Remember: the section numbers included with the headings of the sample sections in this text are intended to be in context with a report not included in this text; therefore, these section numbers have no relevance to this text or to other sample sections in this text.)

5.0 Methodology

The live load at each joint of the truss is determined by combining the maximum unidirectional seismic load with the wind load. For the first iteration of the analysis, the weight of each member is conservatively assumed to be 2.5 times the live load and is distributed equally to each end of that member.

The method of joints is used to determine the force in each member of the truss.

The size of each member is optimized to reduce weight by reiterating the weight and design load.

Critique of the Sample Discussion of Methodology

The first paragraph discusses how the external loads for the structure are combined for the analysis. The second paragraph discusses the theory used to distribute the loads for the design of the members. The third paragraph discusses how the members are designed.

The discussion includes specifics of the analysis ("unidirectional seismic load with the wind load" and "conservatively assumed to be 2.5 times the design load") only as they pertain to the methodology. The criteria for compliance with the customer's requirements are not discussed.

THEORY

A discussion of the theory, including equations (when applicable) and details, should be presented when the readers need it to understand the analysis or calculations presented in the report. When the readers are likely to be familiar with the theory, it is sufficient to include this discussion in the discussion of methodology section of the report.

The entire theory is included in the body of the report only when it can be easily presented. Otherwise, the theory is included in the appendix (see Chapter 6) or is referred to in the references or bibliography.

The theory discussed in the following sample is probably known only by mechanical metallurgists. The details are explained in easy-to-understand language.

Theory

For a slender test specimen, small metallurgical grain size of a metal results in low ductility and high strength because slippage of grains is restricted by the presence and resistance of adjacent grains interfering with the slip planes.

Large metallurgical grain size of a metal results in high ductility and low strength because the centers of adjacent grains are spaced farther apart and the resistance of grains along a plane at an angle to the direction of tension is reduced.

Critique of the Sample Theory

The discussion explains the theory for diametrically opposed conditions (i.e., "small metallurgical grain size" and "large metallurgical grain size"). Each condition is effectively explained using cause and effect (demonstrated by the use of the phrases *results in* and *because*).

DESIGN PARAMETERS

Design parameters specify the load and usage criteria and the physical and performance characteristics used to design a project or system, as shown here:

- A client's specification as the primary source of requirements for a contractor or supplier of a product or service (hereafter referred to as supplier) to comply with.

- An analysis or test report written by a contractor or supplier to demonstrate compliance with a client's requirements. A summary of these parameters informs the reader of the client's requirements.

These parameters must be clearly written and unambiguous. They are usually included in a list. Ordinarily, in a list of items included in a report, the items of greater importance are included before the details. In this instance, the parameters that may affect the overall design are listed before those that affect only a portion of the design.

The following sample design parameters specify the performance and physical characteristics of a vehicle, rather than how it will be designed or built (the subject of the technical requirements).

6.0 Design Parameters

6.1 The acceleration time from 0 to 60 mph shall not exceed 6.9 seconds.

6.2 The maximum speed shall be a minimum of 145 mph, but shall not exceed 165 mph.

6.3 The braking distance at 70 mph shall not exceed 165 feet.

6.4 The curb weight shall not exceed 3200 lbs.

6.5 The minimum turning radius shall not exceed 35 feet.

6.6 The fuel tank capacity shall be a minimum of 16 gallons but shall not exceed 18.5 gallons.

Critique of the Sample Design Parameters

Each design parameter is clearly separated from the others. The format facilitates easy understanding. The list begins with the more important parameters that may affect the overall design (i.e., acceleration and maximum speed are the primary considerations to the designer) and continues with the parameters of the details that affect only a portion of the design (i.e., turning radius and fuel tank capacity are items that can be designed after other requirements have been complied with).

The word *shall* is used rather than *will* or *must* in the statement of each of these parameters. These words are interpreted to have the following meanings in specifications and other legal documents:

- *Shall* is used to *determine* that an event will occur.

The students *shall* be rewarded for their good deeds.
[Although the receipt of a reward is certain, the bestower is not.]

Because the drafter of the specification is the client, who can commit to an obligation only for itself, and *shall* does not identify the actor, *shall* is interpreted to be a requirement that the client (the drafter) does not intend to provide; therefore, the contractor or supplier becomes legally responsible for complying with that requirement.

- *Will* is used as *certainty* that an event will occur.

The students *will* be rewarded for their good deeds.
[The receipt of a reward, as well as the bestower, are certain.]

Will is interpreted to be a provision or condition that the client (drafter) agrees to provide for the contractor or supplier.

- *Must* is a *command* for an event to occur but lacks certainty that the event will occur.

The students *must* be rewarded for their good deeds.
[Neither the receipt of a reward nor the bestower are certain.]

Must is not ordinarily used in specifications or other legal documents, because it is uncertain that the event will occur.

TECHNICAL REQUIREMENTS

In a specification, the technical requirements control the selection of materials and components and the methods for designing the product by the contractor or supplier.

The design parameters discussed in the previous section specify how the project or system will be used. The technical requirements specify how the project or system will be manufactured or constructed. Because both of these requirements control the design, they are sometimes combined into one section entitled Design Requirements.

The following sample technical requirements specify the materials and quality of construction of a house. The design loads, use, and physical characteristics, ordinarily specified in the design parameters section of the report, are not included in this section.

8.0 Technical Requirements

8.1 Beams and columns shall be constructed of Group 1 pine.

8.2 Shear panels shall be constructed of C-D Structural I, 1/2-in. plywood.

8.3 The electrical power system shall accommodate 220-V, 3-phase, 60-Hz alternating current. The minimum power factor shall be 0.95.

8.4 The soil compaction shall be a maximum of 92% of the maximum dry unit weight as determined from Appendix 2, Figure 6.

8.5 The sewer lines shall be fabricated from vitrified clay pipe.

Critique of the Sample Technical Requirements

Similar to the design parameters described earlier, each design parameter is separated from the others in a format that facilitates understanding. Also, the technical requirements that affect overall design are specified before those that affect only the details of design. Again, the word *shall* is used in each of the requirements.

ASSUMPTIONS

When there is an analytical component to the report, it is always based on certain assumptions of the writer. These assumptions are critical to the outcome of the analysis and are clearly specified in a list.

The following sample assumptions section, to be used later in the report, is formatted in a list. Each assumption is identified separately.

Assumptions

The following are the assumptions for the economic feasibility study of the proposed Model 6 standby power generator:

1. The initial purchase price is $10,600.

2. The useful life is 10 years.

3. The salvage value is $4000 at the end of the useful life.

4. The minimum attractive rate of return is 12 percent.

5. Generator operational and maintenance costs are $200 per year and occur at the end of each year.

6. Revenues and operator expenses are not affected by the purchase.

7. The present standby power generator can be salvaged for $1500.

Critique of the Sample Assumptions

Each of the statements is a premise to be used later in the report. The more important assumptions are listed first. Notice, the word *is*, the present tense of *to be*, rather than *will*, the future tense, is used because these assumptions are the basis of this report. See the section on tense in Chapter 3.

These statements state conditions that the writer intends to use as the basis for the analysis. Therefore the word *shall* is not used.

TECHNICAL DESCRIPTION

A technical description is included in reports that discuss a location, structure, process, or product (see Figure 5–2). Because most reports require this information, a technical description is included in most reports. It can be comprehensive for readers with limited prior knowledge, or it can be brief for readers with more extensive prior knowledge. In either case, however, sufficient information should be included for the reader to understand the purpose of the report but not to be overwhelmed with unnecessary details. This section emphasizes comprehensive descriptions. A brief description eliminates details about which the readers are assumed to have prior knowledge.

A technical description is used to help a reader do one of the following:

- Visualize a location (e.g., an airport) or structure (e.g., a ladder).
- Understand a process (a chronological series of events) created by human action (e.g., a manufacturing operation) or nature (e.g., oxidation).

A description of the machinery may not be important to understand some processes such as purifying water. For other processes, such as the manufacture of tires, a description of the machinery used for this process would be helpful.

FIGURE 5–2
Elements of Technical Descriptions

	Technical Description		
	Location/Structure	Process	Product
Purpose	•	•	•
Physical Description	•	•	•
Functional Operation		•	•
Performance Characteristics		•	•

- Visualize the structure and understand the behavioral characteristics of a product with functional components that simultaneously interact with each other to have a single outcome (e.g., an air conditioner that simultaneously moves air as it cools it).

Some descriptions include human action processes that initiate the action of functional components, all of which simultaneously interact with each other to have a single outcome. For example, to explain the operation of an automobile engine, the description must include the physical characteristics of the engine and its functional components, the human process of starting the ignition to initiate movement of the functional components, and the behavioral characteristics of these functional components that simultaneously interact with each other to produce power.

A Technical Description Compared to a Literary Description

A literary description is subjective and includes the opinions of the writer to draw an emotional response from the reader. A sample literary description of a room follows:

> One enters the palatial room through an elegantly carved maple door to reveal the French provincial furniture of another century. The plush beige carpet makes one want to run and dance barefoot.

A technical description, by contrast, is objective and includes the information that the reader needs to comprehend the location, structure, product, or process. A sample technical description of the same room follows:

> The entrance to the 24-ft by 30-ft room is a 36-in. by 80-in. maple door decorated with a carved regal emblem. The floor has a beige nylon carpet with a 1-in. pad. The furniture is French provincial.

The literary description evokes emotions, whereas the technical description describes a scene.

General Guidelines for Writing an Effective Technical Description

The following guidelines (see Figure 5–3) help the reader comprehend your description:

- Use sufficient details so the reader can mentally reconstruct the location, structure, process, or product. Avoid interesting but unnecessary details.
- Include measurements of the five senses (i.e., vision, sound, touch, odor, and taste). Include rates for functions and processes.
- Use precise numbers rather than vague words such as *large*, *fast*, and *very*, which the reader may misinterpret.
- To provide continuity, use spatial transitional words and phrases, such as *above*, *below*, *to the left*, and *adjacent to*, for physical descriptions. Also, use

FIGURE 5–3
Guidelines for Writing Technical
Descriptions

A Technical Description Includes
• Details
• Measurements
• Precise numbers
• Transitional words
• Analogies
• Cause and effect (process)
• Graphics

chronological transitional words, such as, *first, second, next, last,* and *after,* for process and product descriptions.

- Use analogies to explain spatial concepts of size and configuration (e.g., "in the shape of a football 8 inches long and 4 inches in diameter"). Also, use comparison to letter forms (e.g., L-shaped) to explain configuration.

- Use words such as *because, therefore,* and *consequently* to emphasize cause and effect for process descriptions.

- Use photographs and drawings with captions to help the reader visualize a location, structure, or product. Emphasize the details of the description with arrows in the photographs and drawings. You can also emphasize the details with references in the text. Refer to Chapter 9 for more information on using graphics in reports.

Although a technical description ordinarily includes an introduction that states the purpose of the location, structure, process, or product, it does not usually include a summary or conclusion, because this description ordinarily leads into other sections of the report.

Technical Description of a Location or Structure

For describing a location or structure, include the following in the order listed:

1. Purpose: a statement of where the location can be found, and the intended use or function of the location or structure. When it is not obvious, the intended user of this location or structure is indicated.

2. Physical description: the size, shape, location, and orientation of components, material, color, weight, texture, markings, and any other pertinent physical characteristics. First, the location or object as a whole is described. Then, the details of its components are developed.

The following sample technical description of a sidewalk repair emphasizes configuration and size to help a supervisor visualize it.

Repair Description

The site of the sidewalk repair in front of 14 Tahquitz Road in Orange, California, can be seen in Figure 1. This site is typical of a 5-foot-wide sidewalk between an 80-foot-wide residential property and an 8-foot-wide parkway that includes occasional olive trees.

Concrete slabs positioned along the centerline of the sidewalk are 3 feet by 3 feet. Concrete slabs, 1 foot wide by 3 feet long, are on either side of these centerline slabs. The sidewalk is level in the direction of travel and has a downslope of 1/8 inch per foot toward the parkway. Expansion cracks, 1/4 inch wide, separate all slabs (see Figure 2).

One 10-inch-diameter olive tree is planted in front of this house on the centerline of the 8-foot-wide parkway, directly in front of the living room window. The roots of this tree have uplifted the closest 1-foot by 3-foot slab and the 3-foot by 3-foot slab by approximately 1 1/4 inches above the edges of the adjacent slabs. These two slabs, because of this uplifting, have a slope of 3/8 inch per foot. There are no cracks in these slabs (see Figure 2).

To comply with our policy of providing safe sidewalks for pedestrians, the uplifted edges of these two slabs were ground even with the adjacent slabs to eliminate the tripping edge. These ground areas extend approximately 10 inches in the direction of travel, where they are gently contoured into the existing surfaces (see Figure 3).

Critique of the Sample Technical Description of a Location

The description begins with the location of the repair. A general description of the layout of the area takes advantage of prior knowledge of the supervisor by stating "This site is typical of . . ." which encourages the supervisor to rely on previous experience in visualizing this site. Note that the orientation of the street (e.g., east–west) or the side of the street that the sidewalk is on (e.g., south) is not indicated, because this information, while relevant for certain situations, is not relevant to understand the cause of the damage or the repair.

The second paragraph discusses the configuration and details of the sidewalk before any damage occurred. The most important details are discussed first.

The third paragraph discusses the cause of the damage and the changes that have occurred. It also discusses the damage that did not occur (i.e., "no cracks") but is ordinarily of concern, to inform the supervisor that this item was not overlooked. The fourth paragraph discusses the repairs that were performed.

This description refers to three figures (these do not appear in this text) that are referenced sequentially (i.e., the first reference is to Figure 1, the second is to Figure 2, etc.). These figures must appear in the text in the same order in which they are referenced. Also, appropriately, these figures supplement, rather than replace, the information presented in the text; the reader should be able to visualize this location and repair even without the aid of these figures.

Technical Description of a Process

For describing a process, include the following in the order listed:

1. Purpose or source: a statement of the intended use or function of a process created by human action, or the cause of a natural process.

2. Physical description: a physical description of the equipment required for a human action process or the elements that cause a natural process. This physical description is usually more important for human action processes than for natural processes and should be included only when it will help the reader understand this process.

3. Functional operation: a chronological functional description (how it works or occurs) of the process. Include the components (e.g., gears, cranks, microchips, chemicals) for a human action process, or the elements (e.g., water, enzymes) to make the process occur for a natural process.

The following sample technical description of this natural process (which is easily simulated in a laboratory) emphasizes the concept of the process rather than the physical characteristics of it.

10. Description of the Process

Galvanic corrosion occurs when electrical contact occurs between two electrodes—an anode (the electrode supplying the electrons from an external circuit) and a cathode (the electrode receiving the electrons from an external circuit)—in a cell filled with water.

Electrons are released from the anode and flow to the cathode because of the greater electric potential at the anode. The excess electrons at the cathode upset the chemical equilibrium of the system, which liberates hydrogen at the cathode from the hydrogen ions in the water.

This chemical imbalance causes additional electrons to be removed from the anode, which causes a further imbalance of the system. This spontaneous reaction corrodes the anode metal and continues to produce and release hydrogen at the cathode. The excess hydrogen at the cathode plates out in the form of rust.

Critique of the Sample Technical Description of a Process

The description begins with discussion of the elements and the environment necessary for the process. The second paragraph discusses the beginning of the process. The description ends with a discussion of the continuation of the process and the end result.

Because each step (action) must precede the following step (action), the steps (actions) are presented chronologically, and cause and effect are emphasized.

Chronological transition words are also emphasized, but spatial transition words are conspicuously absent in this process description. Because the purpose of this description is to help the reader understand the process, not re-create the experiment, it is not necessary to describe the physical characteristics of the apparatus, because they do not contribute to this understanding.

Words that the reader may not be familiar with are defined in parentheses. This prevents any misunderstanding of terms used in the description. Including the nontechnical word *rust* helps relate to the reader the effect of this process.

Technical Description of a Functional Product

The description of a functional product with simultaneously functioning components combines some of the techniques for describing a structure with those of a human action process. However, because these components function simultaneously rather than chronologically, the components and their functions are presented in the order of their importance to the single outcome.

For describing a product with simultaneous functional components, include the following in the order listed:

1. **Purpose:** a statement of the intended use or function of a functional product. When it is not obvious, the intended user of this product is indicated.

2. **Physical description:** the size, shape, location, and orientation of components, material, color, weight, texture, markings, and any other pertinent physical characteristics. First, the object as a whole is described. Then, the details of its components are developed in the order of their importance with respect to the single outcome.

3. **Performance characteristics:** a performance description (*what* it does) of the item, such as speed, distance, power, rate of production, and minimum and maximum limits.

4. **Functional operation:** a functional description (*how* it works) of the operation and the components necessary to this operation (e.g., gears, cranks, microchips, chemicals). This discussion begins with the major function of this product and then discusses its details.

Because the performance characteristics depend on the functional operation, it is often convenient to discuss these together.

The sample following technical description, with figures, of a mechanical device includes sufficient details to visualize and understand its operation.

8.0 Description

The Model C-20 lift-gate assembly is a device that is permanently attached to the rear end of a standard size cargo bed (90 inches wide and 38 1/2 inches above the ground) of a truck. It is used to lift cargo from ground level onto the cargo bed.

An electrohydraulic control and drive system operates the lift-gate assembly, which consists of a deck assembly and an extension ramp attached to a pair of parallel arms mounted to the frame of the truck. A hinged platform of the deck assembly is 24 inches long by 72 inches wide and has a 12-inch-long, hinged extension ramp for loading. A 10-inch-long bed extension [see Figure 1] is made of steel. The hinged platform, extension ramp, and bed extension have a raised lug pattern [see the arrow in Figure 5–4] for traction.

A control arm is mounted at an angle of 30° to the horizontal on the right side of the 10-inch-long bed extension [see Figure 5–4]. To lower the lift-gate assembly, this control arm is rotated 55° counterclockwise (i.e., down) with a force of approximately 18 pounds. This mechanically opens a release (check) valve, which releases the fluid from the hydraulic cylinder and causes the piston to extend. Travel from the fully up to the fully down position requires 9 seconds and is noiseless. To raise the lift-gate assembly, the control arm is rotated 45° clockwise (i.e., up) with a force of approximately 18 pounds. This actuates a motor powered by a 12-volt battery that drives a hydraulic pump, which retracts the piston into the hydraulic cylinder. Travel from the fully down position to the fully up position requires 7 seconds. There is no gap

FIGURE 5–4
Photograph of a Lift-Gate
Assembly Illustrating a Technical
Description

between the lift-gate assembly and the bed extension when the platform is in the fully up position.

Critique of the Sample Technical Description of a Functional Product

The first paragraph describes the purpose of the lift-gate. The second paragraph includes the source of its power and its physical characteristics. The description discusses the general characteristics before the specific ones.

The third paragraph integrates the performance characteristics and the mechanics of its operation. The discussion is in sequential order: It begins with the operation to lower the lift-gate and ends with the lift-gate returned to the fully up position. Cause and effect are emphasized in this paragraph.

The description uses spatial and chronological transition words to help the reader understand that

- The lift-gate is an assembly, rather than a set of unrelated mechanical components.
- The function of the assembly relies on the sequential functioning of its component parts.

Citing a figure that follows the text is effective for helping the reader visualize the assembly and its operation. Using arrows in the photographs highlights the features discussed in the text (such as shown in Figure 5–4).

INSTRUCTIONS AND PROCEDURES

Performing a task, such as assembling or repairing equipment, performing a test, or driving to a destination, requires a sequence of steps to be performed. This sequence of steps is a set of instructions or procedure.

Instructions and procedures have identical tones and formats. Instructions, however, tell users the method to perform an open-ended task that has no specific criteria for accomplishment (e.g., applying for a loan), and procedures tell users the method for obtaining a specific measurement or result (e.g., test data). Instructions are ordinarily used in assembly and maintenance manuals, brochures, and business correspondence, whereas procedures are ordinarily used in laboratory tests.

Frequently, in instructions, the accuracy of results obtained from any one step is not critical to the results obtained in subsequent steps (although nonperformance of one step may prohibit performance of subsequent steps), and any one step may be repeated without necessarily affecting the results obtained in all subsequent steps. For example, the sequence of steps for setting a VCR is ordinarily critical, but inadvertently setting the channel incorrectly does not affect the duration of time for recording, and for most VCRs, the channel can be reset without affecting this duration of time.

In procedures, however, because a specific measurement is desired, repeating any one step will ordinarily affect the results obtained in all subsequent steps. For example, readjusting the water flow in a pump test would necessarily affect the results in all subsequent steps and the final measurement.

This chapter discusses instructions, although the statements in this section also apply to procedures, which are discussed in more detail in Chapter 11, "Laboratory Reports," and Chapter 21, "Experimental and Test Laboratory Reports."

Writing Instructions

Instructions are written in a step-by-step format, with step numbers, using the imperative form. Using graphics and photographs, cited in the text of the instructions, helps the user.

When writing instructions, include the necessary information only to facilitate completing the operation. Use short, clear, and unambiguous statements for each step. Avoid words such as *change* and *adjust*. Rather, use *increase* (*decrease*), and *loosen* (*tighten*). Also, only use terminology with which the user is familiar.

Place any warnings, cautions, or special instructions before the instruction to which they pertain. In the order of decreased risk of injury or damage, they are as follows:

- Danger (often posted on equipment, and occasionally included in instructions) alerts the operator of equipment of a *condition to avoid* or of *potential misuse* during normal operation that could expose the operator to serious injury or damage the equipment:

  DANGER: Do not place hands or fingers under the anvil.

- A warning is an instruction to prevent injury to the operator or damage to the equipment *during normal operation:*

  WARNING: Wear safety glasses when performing Step 6.

- A caution indicates to the operator a precautionary measure that is applicable only *under certain circumstances*, usually to prevent damage to the equipment:

  CAUTION: Do not operate this equipment below ambient temperatures of 32°F.

- A note, different from danger, warning, and caution, may be used to indicate *informational* items only, which help the user understand the purpose of an instruction:

  Thread the screws into the exterior beam. NOTE: This facilitates the joining process when the exterior beam is lifted into place.

  Notes are not used for information related to the safety of the operators or damage to the equipment.

It is common, although incorrect, to omit articles (e.g., *the*) in front of nouns when writing instructions. This practice can lead to ambiguous instructions when a word can be either a verb or noun, depending on its context.

Example

Attach wire to hook on chain.
[This can be intended to mean either of the following:]

Attach the wire to the hook on the chain.
[This instructs the operator to attach the wire to a hook located on the chain.]

Attach the wire to hook on(to) the chain.
[This instructs the operator to hook (or wrap) the wire on the chain. Note: Although this interpretation is not grammatically correct, neither is the instruction that leads to this interpretation.]

Because many users do not read the entire set of instructions before following them, each step should not require the reader to know of any information included in subsequent steps (i.e., a statement concerning the precision required for a series of steps must precede the first step that requires that precision). Also, present the statements in the same sequence in which the events occur.

Example

Confusing: 5.5 Paint the wood surface after hammering the nails into the predrilled holes.
[This may prompt the user to paint the wood surface before hammering the nails into the predrilled holes.]

Improved: 5.5 Hammer the nails into the predrilled holes.

5.6 Paint the wood surface.
[Presenting these as 2 steps eliminates the confusion.]

Each step in the following sample instructions is short and easy to understand.

Instructions for Changing a Flat Tire

1. Remove the jack, its handle, and the spare tire from the trunk.
2. Place the wheel block diagonally opposite the flat tire to prevent the car from rolling when it is blocked up.
3. Remove the wheel cover with the beveled edge of the lug wrench.
4. Using the lug wrench, turn the wheel nuts counterclockwise approximately 1/2 turn.
5. Position the jack under the lift point on the same side of the car as the flat tire.

WARNING: Make sure no one is in the car during the subsequent operations.

6. Insert the jack handle in the end of the jack. Rotate the handle counterclockwise until the frame of the car is approximately 15 inches from the ground.

DANGER: Do not get under the car when it is supported only by the jack.

7. Rotate the wheel nuts counterclockwise and remove them from the studs.
8. Remove the flat tire.
9. Place the spare tire on the lugs.
10. Finger tighten the wheel nuts in the clockwise direction.
11. Lower the car to the ground.
12. Tighten the wheel nuts.
13. Replace the wheel cover.

Critique of the Sample Instructions

Each instruction begins with an imperative word. However, instructions may also begin with a lead-in instruction such as "*When* the paste is dry, lift the assembly from the fixture" or "*If* the resonant frequency is 30 Hz or greater, decrease the amplitude to 0.05 in."

Step 2, concerning safety, has an explanation to help the reader understand why the step is important. Otherwise, explanations are not given.

The warning, addressing the possibility of persons in the car, is an instruction concerning a condition to prevent injury to the operator or damage to the equipment during subsequent steps. The danger statement, addressing getting under the car, describes a misuse that would expose the operator to serious injury.

In these sample instructions, although the sequence of steps is critical for completion of the desired action, the accuracy of results obtained in each step does not affect the results of subsequent steps.

EVALUATION AND ANALYSIS

An evaluation or analysis is a method for obtaining, as a consequence of deductive reasoning, an unknown truth from known data and information. When words, rather than numbers, are used to achieve these results, the method is usually designated as an evaluation. When numbers are used to achieve these results, the method is usually designated as an analysis. Because many technical reports use a combination of words and numbers, but emphasize numbers, the term *analysis* ordinarily includes evaluation as well as analysis.

For reader comprehension, each step of the numerical analysis should be described with words. Numerical calculations are presented in an easy-to-follow format. Sketches, drawings, and photographs are useful supplements to the calculations.

Figure 5–5 presents a sample analysis of a chair leg failure. It is organized for easy understanding.

Critique of the Sample Analysis

The title of the analysis is headed and underlined for easy identification. It begins with the name of the manufacturer of the chair and the model number. The given information is followed by a list of the assumptions, and a free-body diagram of the chair. After the properties of the chair leg are calculated, the stress analysis is performed. The analysis ends with the calculation of the factor of safety (F.S.) and a final declaration of the adequacy ("OK") of the chair leg.

A free-body sketch is located at the beginning of the analysis for immediate attention. Critical dimensions, forces, and components are indicated, but the sketch is not cluttered with unnecessary details. Dimensions are shown to the nearest one-tenth of an inch.

Mathematical calculations are explained with words. Critical answers are emphasized for easy identification. F_c includes a referral to a list of references located elsewhere in the report. The name of the author and the date of analysis appears at the top of each page. The pages, numbered at the bottom, include the

M. Simon
2/18/95

FAILURE ANALYSIS

Atlantic Decorating Dining Room Chair
Model 62N

Fracture of right rear leg, 7.5 in. above floor.
Diameter of leg at fracture: 1.0 in.
Material: red oak

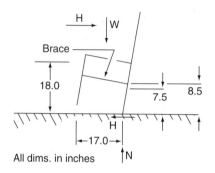

All dims. in inches

Assumptions:

1. Occupant weight, W = 200#
2. Coefficient of friction at floor, μ = 0.75
3. Maximum pushing ability of occupant = 100#
4. All load reacted on rear legs with a 40%/60% load distribution.

Properties of leg at fracture: D = 1.0 in.

A $= \pi D^2/4 = \pi(1.0)^2/4 = 0.785$ in.^{2}

I $= \pi D^4/64 = \pi(1.0)^4/64 = 0.0491$ in.^{4}

S $= I/(D/2) = 0.0491/0.5 = 0.0982$ in.^{3}

W = 200#

H $= 0.75 \times 200 = 150\# > 100\#$

     $\therefore H = 100\#$ max.

For the critical leg:

$P_{COMPR} = 0.60 \times 200 = 120\#$

M $= 0.60 \times 100 \times 7.5 = 450$ in.-lb

f_c $= 120/0.785 + 450/0.0982$

f_c $= \textbf{4880 psi}$

f_s $= 0.60 \times 100/0.785 = \textbf{76 psi}$ (negligible)

F_c = 14,300 psi (Ref.1, p. 6-124)

F.S. = 14,300/4880 = **2.90**

     OK

FIGURE 5–5   Sample Analysis of a Chair Leg Failure

total number of pages (20) in the analysis to alert the reader to an incomplete copy.

When the report is hand written, the handwriting is large, dark, and has well-formed letters for easy reading. It is written in pencil (because ink is not erasable, it is not used for mathematical analysis) on vellum with a light-blue graphic background for excellent quality of reproduction.

Spreadsheets and computer programs are frequently used for analysis. Justification of the validity of the computer program for that application is included in the report. Because of the volume of paper usually generated, a summary sheet of the computer calculations frequently is included in the body of the report, and the computer printout is included in an appendix.

ACCEPTANCE CRITERIA

When laboratory tests are performed to demonstrate the capabilities of the tested items and systems to withstand structural (e.g., forces, abrasion) and environmental (e.g., temperature, humidity) loads during its lifetime, criteria that determine acceptance as a result of exposure to the test environment are specified before the test is begun. These acceptance criteria are usually expressed as the minimum unacceptable tolerances of damage rather than the maximum acceptable tolerances. They must be expressed quantitatively and not subject to interpretation.

Be careful in the use of the word *and*. For example, "Joints shall not rotate more than 1.2° *and* move longitudinally 0.05 in." means that unless both conditions are met, exceeding either tolerance is acceptable. Therefore, when you intend either condition to be unacceptable, you must use the word *or* rather than *and*.

Each item in the following sample acceptance criteria is simply stated and easy to understand.

11.0 Acceptance Criteria

11.1 There shall be no evidence of an abrupt change of reaction at any support during loading or unloading.

11.2 Cracks in fillet welds shall not be longer than 1/16 in. as determined with dye penetrant.

11.3 Deformation of structural members shall not exceed L/180.

11.4 Plastic deformation of structural columns shall not exceed 0.01 percent.

11.5 Surface flatness shall not exceed 1/8 in.

11.6 Paint scratches shall not exceed 2 in. in length.

Critique of the Sample Acceptance Criteria

To avoid misinterpretation, each item states what is unacceptable. For example, if Section 11.6 read "Paint scratches may be up to 2 in. in length," a 3-in. scratch may be incorrectly interpreted to be acceptable because it is not addressed.

KEY CONCEPTS

- Technical descriptions should give the reader a clear picture of the subject of the report.
- The analysis is the critical link between the information and data provided and the conclusions and recommendations.
- The ending of the report should be predictable from the information and data presented in the body of the report.

STUDENT ASSIGNMENT

1. Rewrite the following discussion of methodology for a bridge analysis. Consider grammar, syntax, logic, structure, and content.

   Each load case is examined by first determining the support reactions. Then, by the methods of sections, the internal forces in the members are calculated.

   Once the internal forces are computed, the cross-sectional areas can be obtained.

   Next, the load cases are compared to determine which forces and loads which will be critical for the design of each member. The design loads are then used to determine the final member areas.

   The final step is to compute the composite weight of the structure. This is done by calculating the volume of each member and multiplying it by the given unit weight of the material. These values are tabulated and a 10% allowance is added to give the composite weight of the bridge.

2. Rewrite the following assumptions for a research report. Consider grammar, syntax, logic, structure, and content.

   The following factors may lead to the rusting of steel.

   1. Loads on beams create cracks in the steel causing it to rust.
   2. Exposure to environmental conditions such as air, sun and smog, may cause the steel to rust.
   3. Liquids/chemicals such as acid and water may absorb the chemical coating on the steel beam causing it to lack some of its properties leading it to rust.

3. Rewrite the following design parameters for the colors of a basketball backboard and basket. Consider grammar, syntax, logic, structure, and content.

   8.1 Backboard: The backboard will be painted white. There will be a .5-in. orange stripe 1 ft. up from the backboard. This is the height that the rim needs to be.

   8.2 Rim: The rim and bracket will be painted red.

   8.3 The net will be made of white nylon string.

4. Rewrite the following technical requirements for manufacture of a computer. Consider grammar, syntax, logic, structure, and content.

> **9.1** Keyboard shall be made out of Group 2 Polymer.
>
> **9.2** Hard disk shall be made out of No. 45 metal.
>
> **9.3** Disk drives shall comply with standards listed in computer manual.
>
> **9.4** Memory shall not exceed the circuits maximum capacity.
>
> **9.5** Printer shall be made of same Group 2 Polymer.

5. Rewrite the following instructions for a chemical spill test. Consider grammar, syntax, logic, structure, and content.

> **7.1** Put on rubber gloves and protective goggles.
>
> **7.2** Rinse the beaker in the wash sink and dry it.
>
> **7.3** Pour 100 ml of chlorine bleach into the beaker.
>
> **7.4** Place the beaker in the mechanical arm.
>
> **7.5** Position the beaker over the desk.
>
> **7.6** Press the "SPILL" button.
>
> **7.7** Wearing safety gloves and goggles, after 24 hours carefully wipe the spilled chemical off the desk.
>
> **7.8** Inspect the desk for damage and record any observations.
>
> **7.9** Repeat Steps 7.1 thru 7.8 using 100 ml of coffee, 100 ml of a cola soft drink, 100 ml of ammonia, 100 ml of typing correction fluid, and 100 ml of tap water.

6. Rewrite the following acceptance criteria for a test of a desk. Consider grammar, syntax, logic, structure, and content.

> **9.1** The maximum size of any pieces of wood chipping off shall be 0.5 sq in. or smaller.
>
> **9.2** The maximum deformation of any surface shall be no greater than 3/8 of an inch.
>
> **9.3** Scratches shall be no greater than 1/16 of an inch deep.
>
> **9.4** The maximum abrasion shall be no longer than 1/4 of an inch or wider than 1/16 of an inch.
>
> **9.5** No changes in color or material properties of the surfaces are acceptable.

7. Describe the methodology of one of the following processes to an appropriate audience:
   a. The registration process at your school.
   b. Building a bookcase.
   c. Selecting a major field of study.
   d. Painting your apartment.

8. For a college freshman, develop the theory of one of the following:
   a. Balancing a chemical equation.
   b. Newton's second law ($F = ma$).

     c. Ohm's or Kirchhoff's law(s).
     d. Pythagorean theorem for right triangles.

9. For a design company, develop the design parameters for one of the following designs:
   a. A home entertainment center.
   b. A pair of skis.
   c. A study or den.
   d. A computer.
   e. A vacation home.
   f. A tent.

10. For a design company, develop the technical requirements for one of the designs in Assignment 9.

11. Make a list of assumptions for one of the following situations (review "Types of Technical Reports" in Chapter 1):
    a. As president of your student organization, you write an evaluation of the future needs of the laboratories in your department, to be submitted to the department chair.
    b. As the resident assistant of your dormitory, you write a proposal to your school administration to increase the number of quiet hours in your dormitory.
    c. As treasurer of your fraternity or sorority, you write a feasibility report to purchase ten computers for your members, to be submitted to the officers of your fraternity or sorority.
    d. As a supervisor, you write a proposal to the board of directors to institute flex-time (variable hours) for your employees.
    e. As secretary of your student professional organization, you write an evaluation of a subsidy for your officers to attend a conference.

12. Write a technical description of one of the following:
    a. A room with its furnishings in your apartment or house.
    b. A park.
    c. A commercial building such as a supermarket.
    d. A classroom, laboratory, or other building in your school.
    e. A desk, couch, or dining room set.
    f. The creation of smog.
    g. The four-stroke cycle in an internal combustion engine.
    h. A car-wash.
    i. The registration process at school.
    j. A competitive team sport such as baseball.
    k. A mechanical toy.
    l. A stapler, hole-punch, or mechanical pencil sharpener.
    m. A bicycle, skis, or backpack.
    n. An article of clothing.

13. Write the instructions for one of the following:
         a. Driving from home to school.

      b. Playing a game such as checkers.
      c. Using a washer and dryer.
      d. Planting flowers.
      e. Registering for classes.
      f. Driving a manual transmission automobile.
      g. Bathing a dog.
      h. Drawing a free-body diagram.
      i. Jump-starting a car.
      j. Accelerating onto a busy freeway or turnpike.
      k. Pitching a tent.
      l. Cooking bacon and eggs.
      m. Programming a VCR.

14. For a high school senior, write an evaluation of one of the following:
      a. Your present housing situation.
      b. Your high school preparation for college.
      c. The ability of your vehicle to meet your needs.
      d. A class that you completed within the last year.
      e. Your computer or calculator.

15. For a supplier of products, write a set of acceptance criteria for one of the following to be tested for its anticipated use in its lifetime:
      a. A couch.
      b. An electric power drill.
      c. A watch.
      d. An economy car.
      e. A toy truck or doll.
      f. A college yearbook.
      g. A plastic ballpoint pen.
      h. A lawnmower.
      i. A leather jacket.
      j. A bookcase.
      k. A sleeping bag for backpacking.
      l. A spiral-bound student notebook.
      m. A pair of scissors.
      n. A 6-ft by 8-ft fringed area rug.

CHAPTER 6

The Ending of the Report

The ending of the report is the purpose of the report.

A determination of what has been discovered and recommended courses of action are discussed in the ending of the report. For this reason, it is the most important part of the reporting process. New professionals, inappropriately believing their work is completed when the body of the report is complete and accurate, frequently overlook this importance and write weak endings or omit them altogether. Special emphasis must be placed on writing strong, effective report endings.

When writing conclusions, writers need to shift from deductive thinking (reasoning in which a result follows from the stated premises) used in the body of the report to inductive thinking (deriving general principles from a set of facts). The ending of the report is frequently the most difficult to write because the writer relies on previous experience and judgment rather than the ability to reason. Writers new to the profession should discuss the body of the report with more experienced professionals before writing the ending.

The ending of the report may include the following sections (also see Figure 6–1) shown in the order in which they typically appear:

- Results and conclusions
- Discussion of results
- Recommendations and alternatives
- Bibliography

The appendix is also discussed in this chapter because it follows the report.

FIGURE 6–1
Sections of the Ending of the
Report

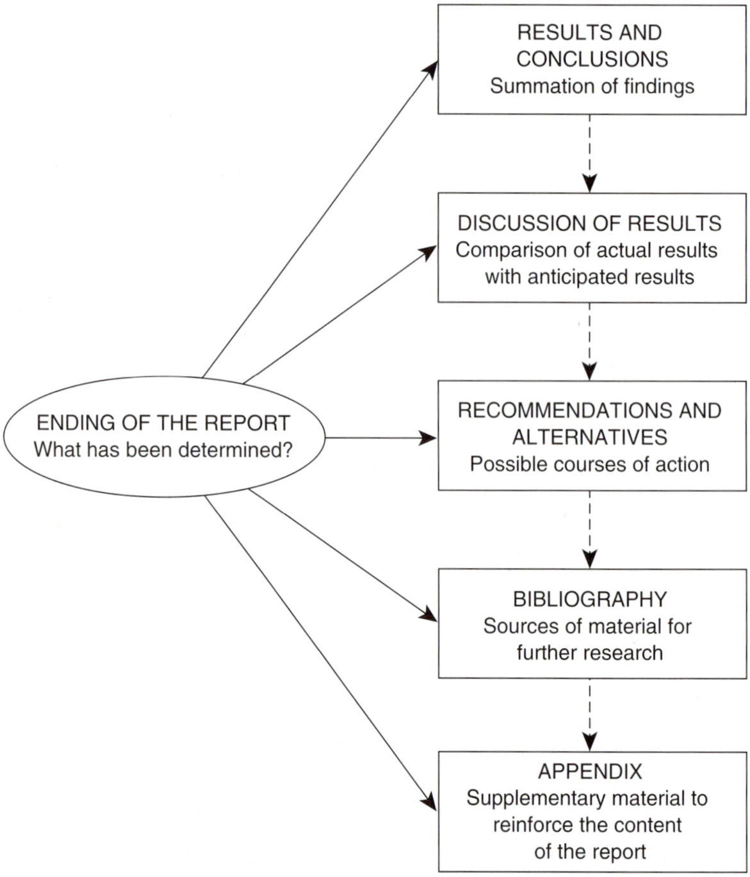

RESULTS AND
CONCLUSIONS
Summation of findings

DISCUSSION OF RESULTS
Comparison of actual results
with anticipated results

ENDING OF THE REPORT
What has been determined?

RECOMMENDATIONS AND
ALTERNATIVES
Possible courses of action

BIBLIOGRAPHY
Sources of material for
further research

APPENDIX
Supplementary material to
reinforce the content
of the report

RESULTS AND CONCLUSIONS

The results are a summation of the findings of an analysis or a test based on the reasoning presented in the body of the report. The conclusions are the inferences drawn from these findings. The results and conclusions are usually included in the same section (as demonstrated in this chapter) because the results from an analysis or test are frequently what the writer hoped to conclude.

When more than one result and conclusion are presented from the evidence, they should be presented in a list, with results preceding conclusions. Because of its importance, this section is sometimes placed near the beginning of a report.

The organization of the following sample results and conclusions easily translates the results of this laboratory test to the immediate conclusion.

11.0 Results and Conclusions

The following are the results from the data and calculations of the efficiency tests:

1. The maximum efficiency is 62% at 450 rpm (see Fig. 4).

2. The specific speed at maximum efficiency is 2.32 (see p. 12).

3. The maximum brake horsepower is 0.60 at 450 rpm (see Fig. 5).

It is therefore concluded that the Model C-80 turbine will convert sufficient kinetic energy at 450 rpm to generate the 0.25 horsepower required to power the Model 361L motor.

Critique of the Sample Results and Conclusions

The results are shown in a list. However, the single conclusion is presented separately as a statement. This statement format facilitates recognizing the difference between the deductive and the inductive reasoning of the writer, while demonstrating the close interrelationship of the conclusion to the results.

Each result includes a reference to the body of the report, demonstrating the relationship of the result to the information in the body of the report.

When there are several conclusions, separate them, and preface the list with a statement similar to "The following are the conclusions of this test determined from the preceding results."

Frequently, a result from the analysis or test is the conclusion. For example, when the purpose of the test in this sample was to determine (1) the maximum efficiency at 450 rpm, (2) the specific speed at maximum efficiency, and (3) the maximum brake horsepower at 450 rpm, these results from the test are also the conclusions, without requiring inductive reasoning to reach them. Therefore, it is common to combine the results and conclusions in the same section.

DISCUSSION OF RESULTS

For an analysis or test, the section discussing results compares the anticipated results with the analytical or test results. This section states whether the anticipated results agree with the analytical or test results. It also discusses any assumptions that were modified for the analysis or any revisions in the test procedure.

When an analysis or test is used to predict behavior or capability, and the behavior or capability is other than predicted, this section may be used to explore the validity for that application of the analysis or test procedure.

The readers need sufficient details to understand the explanation. In the following sample discussion of results, the writer did not anticipate the problem that was encountered during the experiment.

9.0 Discussion

The experimental curves of torque (in.-lb) vs. angle of twist (ϕ) for the cast iron and aluminum specimens do not demonstrate well-defined proportional limits or yield points (see Figure 2). Therefore, the experimental torsional modulii of rigidity, G (psi), cannot be accurately determined. This mechanical behavior is typical of brittle materials. Therefore, to determine the experimental modulii of rigidity, a 0.02%

offset is drawn to determine the yield point of each specimen. The torsional modulus of elasticity is then assumed to be the slope of the line extended from the origin to the yield point.

The experimental values of the torsional modulii of elasticity are 19% and 17% higher than the published values, respectively, for the cast iron and aluminum specimens (Ref. 6, p. 83). These values are significantly greater than the ordinary experimental deviation. However, as the extreme fibers of each specimen reach the elastic limit, a combination of increasing plastic deformation of the outer fibers and elastic deformation of the inner fibers react to the applied torsion (see Figure 3). When plastic deformation of the outer fibers begins, the polar moment of inertia, J (in.^{4}), increases as outer fibers become more effective in resisting torsion. This additional polar moment of inertia from the plastic deformation increases the torsional stiffness of the specimen and the apparent elastic torsional modulus of rigidity.

To prevent this phenomenon, the torque should be limited to values that are significantly below the proportional limit.

Critique of the Sample Discussion of Results

The first paragraph discusses the revised procedure used to determine the experimental values after a problem is encountered. It states that this problem is typical for the type of material tested.

The second paragraph reveals that the experimental values are higher than anticipated. Then, the writer explains, in detail, the mechanical behavior (rather than the revised procedure) that is believed to be responsible for the higher values.

The third paragraph discusses a procedure to prevent this undesirable mechanical behavior in the future.

RECOMMENDATIONS AND ALTERNATIVES

Recommendations and alternatives are possible courses of action the reader can take to achieve the result desired. The writer is not responsible for implementing these recommendations and alternatives, or even justifying their feasibilities; however, they should be practicable.

Although recommendations are very similar to alternatives, there is a difference in the scope of work required to implement them. Recommendations either accept the item or proposal that is the subject of the report or include suggestions that the reader can implement to *improve* the item or proposal. Alternatives are suggestions that the reader can implement to *replace* the item or proposal that is the subject of the report to achieve the same result.

Recommendations and alternatives are a consequence of the conclusions in the report and are determined by inductive thinking. Therefore, the recommendations and alternatives section immediately follows the section that includes the conclusions. This section, similar to results and conclusions, also may appear at the beginning of the report because of its importance.

Because of the close relationship between recommendations and alternatives, they are usually placed in the same section of the report. For the convenience of the reader, however, they should be listed separately, with recommendations preceding alternatives.

The organization of the following recommendations and alternatives clearly separates the recommendations from the alternatives.

13. Recommendations and Alternatives

The short- and long-term benefits of this project are significantly greater than the adverse effects on the environment; therefore, Environmental Technology, Inc., recommends that the proposal for this project be approved.

The following recommendations are made to reduce the impact on the environment:

1. Limit the height of the office building to four stories to minimize the traffic impact caused by vehicles entering and exiting the garage.
2. Relocate the garage access on Central Avenue to 24th Street to eliminate the traffic impact on Central Avenue.
3. Relocate the entertainment facility to the interior of the building to eliminate potential disturbances to the surrounding area.

The alternatives considered do not offer comparable benefits.

Critique of the Sample Recommendations and Alternatives

The section begins with a statement clearly recommending the proposed course of action. Recommendations are made to minimize the impact on the environment. The last statement in the section indicates that the writer considered alternatives to the proposed course of action but that these alternatives did not offer comparable benefits.

The format presented in this sample is similar to the format in the results and conclusions section.

BIBLIOGRAPHY

The bibliography (different from references, which are discussed in Chapter 7) is a list of sources of material for further research by the reader. A bibliography is commonly included in research and nontechnical reports to explain scientific and social theory. Unlike the references, the bibliography is not required for the reader to fully understand the report, and it is not cited in the body of the report.

The citations for the bibliography are alphanumeric and may include page numbers after the name of the publisher. For example, a citation for a book in a bibliography may read:

McCormac, E., *Surveying Fundamentals*, 2nd Ed., 1988, New York, McGraw-Hill, 231–243.

A citation for a journal article may read:

Smith, B., "Comparison Between Geochemical and Biological Estimates of Subsurface Microbial Activities." *Microbial Ecology* 2(1988): 20–25.

When the report is written for a technical or professional journal, this journal usually has its own format to be followed for citations.

APPENDIX

An appendix includes supplementary material used to reinforce the material in the report, for example, extensive theory or catalogs.

Reports frequently have separate alphabetical or numerical appendices, such as Appendix A, Appendix B or Appendix 1, Appendix 2 for the different supplementary materials contained in it.

- When any appendix is longer than two or three pages, a title page introduces the appendices section, and a separate title page precedes each appendix with the title Appendix A (or B) or Appendix 1 (or 2). Samples of appendix title pages are shown in the reports illustrated later in the book.
- Otherwise, head each appendix on the top of its first page with the title Appendix A (or B) or Appendix 1 (or 2).

Page numbers in an appendix are preceded by the letter designation of the appendix. For example, designate the third page in Appendix C as C3.

KEY CONCEPTS

- The results and conclusions should be based on the information and data presented in the body of the report.
- The discussion of results is the primary source of information when behavior or capability are other than predicted.
- Recommendations and alternatives require inductive, rather than deductive, thinking.
- Supplementary material in the appendix may be used to reinforce the information in the report.

STUDENT ASSIGNMENT

1. Rewrite the following conclusion for the evaluation of a desk. Consider grammar, syntax, logic, structure, and content.

An alternative to any desk should always be considered. The desk should satisfy the needs of the student. It should be minimal in cost, practical, and easy to transport. After evaluating my desk, I identified several issues. Because of the shape of the desk, an adjustable drafting board is recommended for drafting. A file cabinet should be provided for storage space. The desk should resist heavy loads such as from a computer. Furthermore, because of the dimensions of the desktop, the desk is suited for large projects. Overall, the desk serves its purpose, and it is a pleasant place to work.

2. Rewrite the following discussion of results for the evaluation of a new computer program. Consider grammar, syntax, logic, structure, and content.

The computer program generates the stiffness matrix in two ways—the inverse method and the direct method.

The inverse method produced a perfectly symmetrical stiffness matrix. This makes perfect sense because the steel frame is perfectly symmetrical in its properties, and the applied loads are also perfectly symmetrical. The direct method produced a slightly asymmetrical stiffness matrix. The elements k_{ij} and k_{ji} were not exactly equal and differed by an average of 0.95% and by a maximum of 1.90%. These are acceptable error ranges and are effectively compensated for by the safety factors that are to be used later on in the design calculations.

The direct method is more sensitive to the geometric shape of the structure to be analyzed and any asymmetry, whether it exists between the left and right sides or the top and bottom, will cause roundoff errors which in turn will produce an asymmetrical matrix. The best remedy is just to simplify the structure to be modeled and to rely more heavily on the matrix produced by the first method.

3. Rewrite the following recommendations and alternatives for an environmental impact report concerning a pond. Consider grammar, syntax, logic, structure, and content.

The picnic area location could be designed to eliminate the need to relocate the willow tree and prevent damage to its health. This could be done by relocating the picnic area or by incorporating the tree into its design. This would also eliminate the effect on the endangered Spotted Owl.

The proposed quantity of trout should be reduced to fifty and the population should be regularly monitored and maintained. The duck population should also be regularly monitored and maintained. This will mitigate the long term effect on traffic safety and the level of service on Second Street.

The pond bottom should be lined with clay to inhibit the groundwater infiltration. This will mitigate any effects on the water quality.

4. Write results and conclusions for one of the following situations (review "Types of Technical Reports" in Chapter 1):

   a. As treasurer of your student professional organization, you organized and managed a fundraiser (e.g., a carwash or pancake breakfast) and reported it in the student newspaper.

   b. As house manager of your fraternity or sorority, you managed the food service for an academic year and evaluated it for your successor.

   c. As a computer expert, you evaluated a new computer program for your supervisor at your job.

    d. As a student, you evaluated one of the following for the administration: a new registration procedure, class, laboratory, building, or recreation facility.

    e. For the coach of your athletic team, you evaluated the effect of new equipment on your team's record.

5. Write the discussion of results for one of the situations in Assignment 4.

6. Write the recommendations and alternatives for one of the situations in Assignment 4.

CHAPTER 7

The Beginning of the Report

*The beginning of the report prepares the reader to understand
the body of the report.*

The beginning of the report introduces the body of the report. It includes the background information needed to comprehend the material discussed in the report.

The beginning of the report may include the following sections (see Figure 7–1) shown in the order in which they typically appear:

- Abstract
- Summary
- Introduction
- Background
- Scope
- Applicable documents
- Nomenclature and Definitions
- Symbols
- References

Although some writers place the summary at the end of the report, most writers prefer to place it at the beginning because, except for conclusions and recommendations, it is the most important section of the report for readers not intending to read the entire report.

Writers usually write the beginning of the report after completing the other parts of the report, because the subject matter included in it is ordinarily based on a knowledge and understanding of the information, facts, and data discussed in the body and the ending of the report.

The first three sections in this chapter, abstract, summary, and introduction, discuss sections of technical reports that have similarities. A summary of these sections and their distinctions follows (see Figure 7–2):

FIGURE 7–1
Sections of the Beginning of the
Report

ABSTRACT
Discusses the contents
of the report

SUMMARY
A concise statement
of the report

INTRODUCTION
Discusses the purpose
of the report and the
source of authorization

BACKGROUND
A history to give an understanding
of the purpose of the report

SCOPE
Discusses the limits of the study
of the subject matter (report),
or addresses the extent of the work
to be completed (specification)

APPLICABLE DOCUMENTS
A list of documents incorporated
into a specification

NOMENCLATURE/DEFINITIONS
A list of terms with
special meanings

SYMBOLS
A list of symbols used in
an analysis or test

REFERENCES
A list of sources of information
required to fully
understand the report

BEGINNING OF
THE REPORT
Introduces the body
of the report

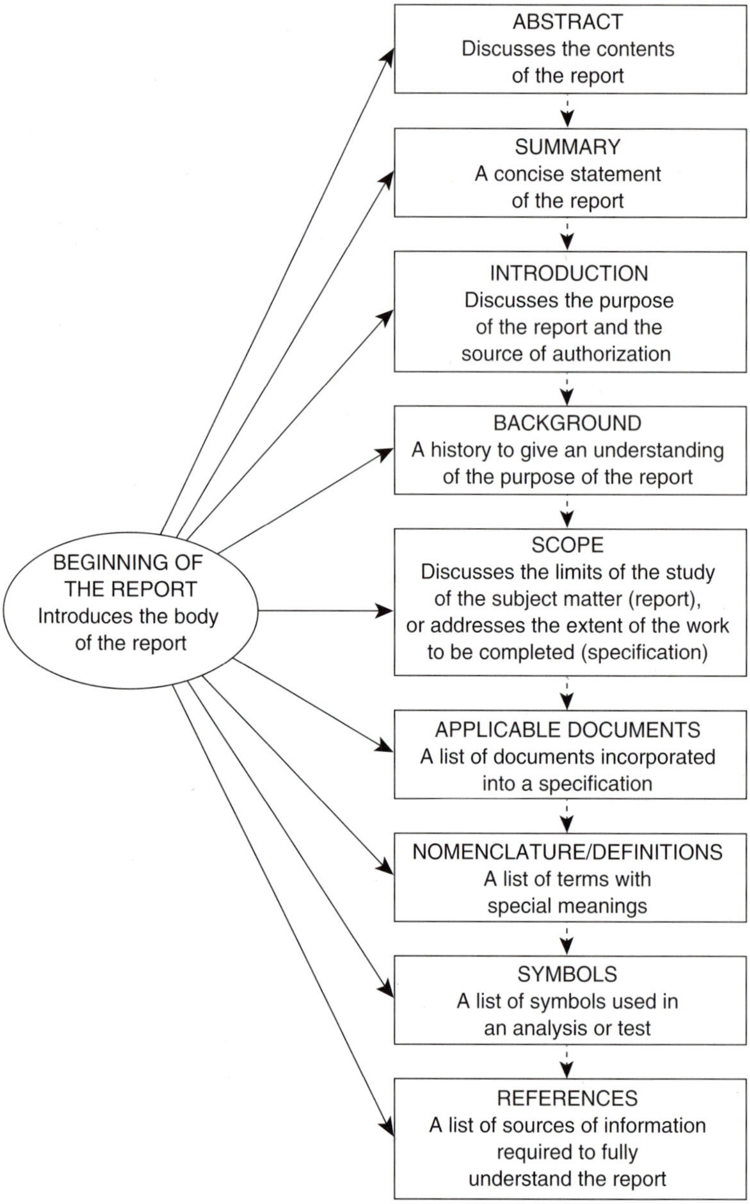

- An abstract tells potential readers the contents of the report.

  A descriptive abstract enables these potential readers to determine when the report pertains to their interests or needs. Conclusions and recommendations are not included.

  An informative abstract summarizes the contents of the report and includes conclusions and recommendations.

FIGURE 7–2
Abstracts, Summaries, and Introductions

	Descriptive Abstract	Informative Abstract	Summary	Introduction
Purpose	•	•	•	•
Facts and data	•	•	•	
Assumptions		•	•	
Procedure		•		
Conclusions		•	•	
Recommendation/alternatives		•	•	
Source of authorization				•

- A summary is an informative abstract but may also include information concerning the procedure or methodology used.
- An introduction tells readers the purpose of the report and its authorization. It does not address the subject matter of the report.

ABSTRACT

An abstract, when included in a report, is always the first section. Because it is not intended to be an integral part of the report, it is often printed in bold type or italics. An abstract is most commonly included in research reports and technical articles. An abstract can be either descriptive or informative.

Descriptive Abstract

A descriptive abstract gives an overview of what is discussed in the report so readers can decide whether the report pertains to their interests. It is typically one or two paragraphs long. Conclusions and recommendations are not included in the abstract.

When an abstract is included in a report, it is generally descriptive rather than informative unless there is a specific requirement for an informative abstract.

In the following descriptive abstract, the writing is clear and concise. It discusses the report information but does not include conclusions or recommendations.

Abstract

Government metallurgists accidentally discovered an improved method for welding steel by adding an alloy to the welding rod. Frequent structural failures in continuous overhead seam welds at utility plants encouraged the federal government to subsidize a research program to develop a welding rod material that would prevent brittle hardening of the parent material. After many tests, an additive was found that not only prevented brittle hardening but also reduced the time required per pass of weld and increased its strength by 15 percent.

Critique of the Sample Descriptive Abstract

This abstract begins with a statement of an event and its history. Even though the advantages resulting from the event are discussed, the methodology and solution for arriving at the event are not revealed. An informative abstract would include this methodology and solution as well as recommendations for future applications.

Because the abstract precedes the report and is not an integral part of it, this sectional heading is not assigned a section number. Section numbers, when used, begin with the first section after the abstract.

Informative Abstract

An informative abstract (sometimes called an "Executive Summary") summarizes the information in the report and includes conclusions and recommendations, ordinarily not included in descriptive abstracts. For readers who are primarily interested in results, the informative abstract replaces reading the report. An informative abstract may contain procedural details that summaries do not include.

For example, if the preceding abstract concerning a metallurgical discovery were to be revised to become an informative abstract, information concerning the type of tests used in this research program, the content of the additives attempted, the additive that accomplished the intended result, and potential applications would be added.

Professional journals frequently require an informative abstract to be included with the article. The length and style for the abstract are usually specified. Frequently, professional societies publish a periodic index of informative abstracts.

Because the content of an informative abstract may be the same as a summary, when an informative abstract is included in a report, ordinarily, a summary section is not. The discussion of informative abstracts is included in the following section, entitled "Report Summary."

REPORT SUMMARY

The summary is an abbreviated form of the most important parts of the report. The following items are usually addressed in the summary: purpose of the report, given facts and data, assumptions, conclusions, and recommendations and alternatives. Details are not included. An informative abstract will include the same items as the summary but also may include information about the procedure or methodology used.

The summary is probably the most frequently read section of the report and therefore should be carefully written to prevent readers from misinterpreting its content. Technical jargon, information, and data should be kept to a minimum because many readers do not have a technical background. Executives often use the summary as a basis for making decisions. Others may review the summary to refresh their memories.

The summary represents the contents of the report and should leave the reader with the same impression as when the report was read in its entirety. It should only include information that is presented in the body of the report. When using building-block structure, a summary belongs near the end of the report, but because it is important to readers who may not read the entire report, it is frequently placed after the abstract.

The information in the report is briefly reported in the following sample summary. It includes a conclusion and recommendation.

1.0 Summary

To relieve overcrowding of the nation's air transportation system, Congress has proposed funding a cross-country railway system for the next decade. The Department of Transportation studies claim that sufficient passengers could be drawn from pleasure travelers, predominantly senior citizens, to make the system self-supporting.

Our investigation concludes otherwise. Although 8 percent of cross-country travelers are senior citizens, a segment growing approximately 2 percent annually, less than 1 in 20 of them judge the railway system convenient enough to replace air travel. This falls short of the numbers required to free the system of federal subsidies.

Consequently, federal funding of this railway is not recommended.

Critique of the Sample Report Summary

The summary begins with a statement of a proposal and its purpose. Facts and data are analyzed to determine the feasibility and to make a recommendation for the proposal. Except for details, all pertinent information is included in the summary.

INTRODUCTION

Unlike reports written in an academic setting, the introduction to a professional report is concerned with the report itself rather than the subject matter of the report. It discusses the function of the report, what it expects to accomplish, and the source of the authorization. The introduction presents material that is germane to readers before they read the body of the report.

The following sample introduction includes a statement concerning the background and purpose of the report and the circumstances under which the report is written.

2.0 Introduction

The state-of-the-art technology for fabricating and finishing steel pipe is rapidly changing. This research report studies the relationship between pipe surface roughness and the capability of a pipe to carry an incompressible fluid under pressure to determine if the added expense of ultra-finish pipe is justified.

> This research and report is to demonstrate compliance with the requirements of Section 6.0 of Contract No. 62-361C with the U.S. Department of Energy.

Critique of the Sample Introduction

The subject matter of the report is mentioned only to explain the purpose of the report. The methodology, results, conclusions, or recommendations are not included. The second paragraph clearly indicates the contractual reason for writing the report. The source of authorization, although not stated, is implied to be from these contractual requirements.

BACKGROUND

A historical background helps the readers to understand the circumstances that lead to the need for the work or project that is the subject of the report. It may discuss changes in technology, economics, or legal or contractual requirements. When the background is brief, it can be included in the introduction section.

Background

> As a result of the bankruptcy of the sole supplier of 9/16-inch brass needle valves, an alternate method of controlling the flow of gas at 40 psig must be developed for Assembly 26-32. This needle valve allowed close control of the flow and was noncorrosive in a corrosive environment.

Critique of the Sample Background

The background clearly states the reason for developing a new method or part and the function of the previously used part. This explanation clarifies the importance of the project.

SCOPE

The scope of a *report other than a specification* discusses the limits of the study of the subject matter that is addressed in the report. Also, it includes any items that are not addressed specifically in the report or are addressed in other reports.

The scope of a *specification* (a set of contractual requirements for compliance) addresses the extent of the work or project to be completed. Also, it addresses any work that is specifically excluded.

The following sample scope for a report other than a specification identifies the limits of what is analyzed in the report (a structure at a specific address) and the basis for this analysis (the Uniform Building Code and the American Institute of Steel Construction). It includes what is not analyzed in the report.

2.0 Scope

This report analyzes the structure of 94 Main Street in Bigtown, Nebraska, for the loads specified in the 1994 Uniform Building Code. Material properties are as specified by the American Institute of Steel Construction. Deflections that are due to design loads are not the subject of this analysis.

Critique of the Sample Report Scope

The subject of the report is clearly defined with a street address. The basis for the analysis is stated, without details, to alert the reader of the limitations of the analysis.

Deflections (movement of structural members), which are very often included in structural analyses, are specifically excluded in the report scope. This informs the reader that calculation of deflections was intentionally omitted rather than an oversight.

In the following sample scope from a specification, the work that is required to be supplied is addressed, but a limitation on the contractor responsibility is included.

2.0 Scope

The contractor shall supply twelve (12) printed circuit boards and twelve (12) digital controllers as specified herein. All components shall be assembled and packaged in packaging supplied by the owner.

Critique of the Sample Specification Scope

The contractual requirements for the circuit boards and controllers are clearly stated. The limitation for the responsibility of packaging (i.e., "supplied by the owner") is included.

This scope, a statement of requirements to be complied with, is written in the imperative tense. Also, because the specification is a legal document, the number of units to be supplied is specified using words followed by digits in parentheses, for example, "twelve (12)."

APPLICABLE DOCUMENTS

The list of applicable documents in a specification includes codes, standards, and other specifications required for compliance with the subject specification. They are listed in alphanumeric order to facilitate retrieval from the reference library.

Using applicable documents allows the writer of the specification to incorporate previously written requirements or data with a short statement only. For example, a paragraph in the body of the subject specification might read "Dimensioning of structural members shall comply with Section 4.0 of ANSI Y14.5." ANSI Y14.5 is included in the list of applicable documents to alert the reader (user) of the subject specification that ANSI Y14.5 will be needed for compliance and must be made available.

In the following sample list of applicable documents, the items are numbered and in alphanumeric order for the convenience of the user and the librarian.

3.0 Applicable Documents

1. DOT-HS-801, Vol. I, Greene, J. "Occupant Survivability in Lateral Collisions" January 1976.
2. *Federal Register*, Vol. 53, No. 161, "Advanced Notice of Proposed Rulemaking to Revise Federal Motor Vehicle Safety Standard No. 214—Side Impact Protection" January 1988.
3. SAE Report No. 830459, Partyka, S., & Rezabek, S., "Occupant Injury Patterns in Side Impacts—A Coordinated Industry/Government Accident Data Analysis."
4. SAE Report No. 890386, Daniel, R. P., "Biomechanical Design Considerations for Side Impact."

NOMENCLATURE OR DEFINITIONS

The nomenclature or definitions section includes a list of terms used in the report that have special meanings that readers need to understand.

5.0 Definitions

5.1 Intermittent weld—a 3-inch fillet weld, 12 inches on center (3–12) unless otherwise specified

5.2 Pipe—2-inch Schedule 40 standard steel pipe per ASTM A36

5.3 Drill—boring a hole with a 1/3-HP drill rotating at 3600 rpm

Critique of the Sample Definitions

These definitions help to simplify the writing when a standard item or procedure is used repeatedly. For example, whenever the term *pipe* is used in the report, it means "2-inch Schedule 40 standard steel pipe per ASTM A36."

SYMBOLS

In a report that has an analytical or test component, a list of symbols is usually included unless the symbols are standard ones that all knowledgeable readers in that discipline or specialty should understand. Symbols are the nomenclature of physical and mathematical quantities and concepts discussed in a report.

4.0 Symbols

S_p = proportional elastic limit

S_Y = yield point

S_t = ultimate tensile strength

S_c = ultimate compression strength

μ = Poisson's ratio

Critique of the Sample List of Symbols

The subscripts modify the basic symbol S, which means strength.

Although it is not recommended, writers sometimes define their own symbols for physical and mathematical quantities and concepts ordinarily represented by standard symbols.

REFERENCES

References (different from a bibliography, discussed in Chapter 6) are sources of information that are required so the reader can fully understand the body of the report. These references are ordinarily used for lesser known scientific and engineering facts and equations when the author anticipates these to be beyond the expertise of the reader.

The items included in a list of references are sequentially numbered in the same order that they appear in the body of the report. Each reference includes the author, title and edition, date, and the publisher's location and name.

This system eliminates multiple citations of the same source of information in the body of the report. Without a reference list, the reference in the body of the text would read "See Timoshenko, S., & Goodier, J., *Theory of Elasticity*, 3rd Ed., 1969, New York, McGraw-Hill, p. 245." However, with this list, the citation is shortened to "See Ref. 1, p. 245" or "(Ref. 1, p. 245)."

Note the following distinctions:

- Specifications include a list of applicable documents (discussed earlier) for compliance by the supplier or provider of products or services. The reader must obtain these documents to determine the full scope of the work to be performed.

- Reports include a list of references (discussed in this section) to help the reader understand the information in the report. Usually, reports intended for publication will include the references at the end of the report rather than at the beginning.

- Reports also may include a bibliography (see Chapter 6) for the reader interested in further research. Bibliographies are most commonly found in research and nontechnical reports as a source of additional information for the interested reader.

- Footnotes (found at the bottom of a page) and endnotes (found at the end of a chapter or book), include supplementary information and explanations and literary citations for the reader. They can be used to give credibility to a theory or opinion of the writer that might otherwise be deemed overly subjective.

Callouts to footnotes and endnotes are numerically superscripted in the text in the same sequence they appear. The footnotes and endnotes then appear in their respective locations in the same sequence as their callouts. Because footnotes and endnotes are not ordinarily included in technical reports, they are not discussed further in this text.

In the following list of references, the items are presented in the sequential order of their first citation in the text of the report.

References

1. Peurifoy, R., & Ledbetter, W., *Construction Planning, Equipment and Methods*, 4th Ed., 1985, New York, McGraw-Hill.
2. Avallone, E., & Baumeister, T., *Standard Handbook for Mechanical Engineers*, 9th Ed., 1987, New York, McGraw-Hill.
3. Spencer, A., Powell, G., & Hudson, D., *Materials for Construction*, 1982, Reston, VA, Reston.
4. Barrie, D., & Paulson, B., *Professional Construction Management*, 2nd Ed., 1984, New York, McGraw-Hill.

Critique of the Sample References

This method of referencing sources of information (author[s], title, edition, date, and publisher's location and name) is commonly used in the practice of engineering and science. However, this method of citation is not uniformly accepted, and practices in specific fields of technology may differ.

KEY CONCEPTS

- Because the beginning of a report organizes the material for the reader, the beginning sections are usually written after the body and end of the report are complete.
- The abstract is ordinarily included in research reports and technical articles.
- The scope is ordinarily included in any report with limitations on the study of the subject matter of the report and in specifications.

STUDENT ASSIGNMENT

1. Rewrite the following descriptive abstract for a research report. Consider grammar, syntax, logic, structure, and content.

   Oxidation of steel has become a major problem encountered by many engineers and scientists. After many tests were performed, metallurgists at the Michael Institute have come up with a solution on why steel rusts. In this report four tests will be performed to identify the reasons for steel rusting.

2. Rewrite the following proposal summary for a parking structure. Consider grammar, syntax, logic, structure, and content.

> Smith General Contracting Co. proposes a $1,420,000 parking structure. The initial stage covered in the proposal is the construction of a 2-story, concrete parking structure. It will be located on the corner of Main St. and Spielman Road in Drylawn, New York. As the demand for more parking stalls increases, the parking structure will be expandable to a maximum height of six stories.

3. Rewrite the following introduction of an evaluation of food. Consider grammar, syntax, logic, structure, and content.

> As resident assistant of the dormitory, numerous complaints about the food have prompted me to evaluate the food service in the cafeteria. Of course, complaints of cafeteria food are quite common at many schools. However, the number of complaints have been increasing this year. Reports of students getting sick with food poisoning have been quite alarming. This situation should not go on for much longer. I hope my evaluation will be useful in determining measures to be taken to improve food service.

4. Rewrite the following scope written for a procedural test specification of a casing. Consider grammar, syntax, logic, structure, and content.

> Out of the one thousand casings manufactured, 50 casings will be tested for dropping, bumping, wear, abrasion, and liquid/chemical spills. The tests will show that the casings will withstand the stresses compelled by all anticipated use of the stereo equipment during shipment and a seven year life in the home or business.

5. Rewrite the following introduction for a feasibility report concerning a part-time job. Consider grammar, syntax, logic, structure, and content.

> A civil engineering student has the option of purchasing a used car so that she can accept a part-time career-related job off-campus. It would be to her benefit to make such a purchase and take the job being offered.

For Assignments 6 through 10, consider the following situations (review "Types of Technical Reports" in Chapter 1):

a. As secretary of your student professional organization, you write an article for your school newspaper about a regional technical competition in which this organization participated.

b. As president of your student professional organization, you write a proposal to the chair of your department to include a technical procedure into a 4-unit class required for students in your major.

c. As the resident assistant of your dormitory, you write an evaluation to the director of housing services of the food provided by the food service in your dormitory.

d. As treasurer of your fraternity or sorority, you write a feasibility report to its national headquarters concerning the purchase of a new house closer to campus.

e. As a newly promoted supervisor, you write a proposal for management to purchase a computer for you to use for your administrative responsibilities.

Supplement the preceding situations with details that create purposes and contexts and write the following (approximately one paragraph each):

6. A descriptive abstract for Situation A.
7. A summary for one of the situations.
8. An introduction for one of the situations.
9. A scope for one of the situations.
10. A background for one of the situations.

CHAPTER 8

The Cover Materials

The cover materials set the tone for reading the report.

The cover materials envelope the report and give readers a preview of its contents. These materials include the following shown in the order in which they appear:

- Title page
- Cover letter or memo
- Invoice
- Letter of transmittal
- Table of contents

Readers receive their first impression from these materials; therefore, the cover materials should appear professional.

Many organizations use standard formats for the cover materials to ensure uniformity. Often, a nontechnical officer of the organization prepares the cover letter or memo, invoice, and letter of transmittal. These materials are ordinarily prepared after the report's completion because they address its contents.

TITLE PAGE

The reader sees the title page before the contents of the report; therefore, it should have a neat and organized appearance. The title page should clearly indicate the subject title in the upper half of the title page. A report number, when one has been assigned, appears directly below the title. The person or organization receiving the report, the person or organization writing the report, the date of the submittal, and the contract number, when one has been assigned, appear in the lower half of the title page.

Some organizations require that reports submitted to it from other organizations use a standard format. When this is the case, all information appears clearly and legibly.

The title of the report specifically identifies both the report's function (what the report does) and subject (what the report concerns).

Examples

Vague: Motor Analysis

Clear: Performance Analysis of Model 26K Three-Phase Motor

Vague: Study of Apartment Building

Clear: Cost Feasibility Study of Oceanview Towers

[Note: a feasibility study can address either cost or technical feasibility; therefore, the appropriate feasibility should be indicated in the title.]

Figure 8–1 illustrates a sample title page that is attractive and informs the reader of important information.

Critique of the Sample Title Page

Information relating to the function and subject, report number, writer, recipient, contract number, and date submitted (important in contract disputes) are clearly visible to the reader in the sample title page in Figure 8–1. The placement of the information is balanced and esthetically pleasing to the reader.

The title clearly identifies the contents of the report. It begins with a functional heading, "Stress Analysis," which tells the reader what the report does. The subject of the report, "ST-42 Aileron Support Structure," immediately follows the functional title, which tells the reader the reported subject. Because both of these are the most important items on the cover page, they are in uppercase letters in the upper half of the page. The report number appears immediately below the title.

Less important items are titled for easy identification and indicated in the lower half of the page.

COVER LETTER OR MEMO

When the recipient of the report is not the official who authorized it, a cover letter (used for interorganizational reports) or memo (used for intraorganizational reports) addressed to the recipient identifies the title, subject, and the writer of the report. Because only the recipient reads the cover letter or memo, social courtesies and informal comments relating to the project are included. An invoice, when one is enclosed, is referenced. Also, the recipient may be thanked for the opportunity to write the report. The cover letter or memo is loosely inserted behind the title page and in front of the invoice, when one is attached, and the letter of transmittal.

The structure and contents for cover letters, memos, and letters of transmittal are discussed in this chapter. See Figure 8–2 for a comparison of cover letters/memos and letters of transmittal.

STRESS ANALYSIS

ST-42 AILERON SUPPORT STRUCTURE
Model 80 Airplane

Report No. 95-560

Date Submitted: May 14, 1995
Contract AF 30L

Prepared by: *Prepared for:*

Structural Mechanics, Inc.   *Wing Manufacturing, Inc.*
Mytown, IS   *Yourtown, BE*

FIGURE 8–1
Sample Title Page

FIGURE 8–2
Cover Letter/Memo and Letter of Transmittal

Cover Letter/Memo	Letter of Transmittal
• Addressed to recipient	• Addressed to authorized client official
• Identifies title	• Identifies title
• Identifies subject	• Identifies subject
• Includes social courtesies	• Includes statement of transmittal
• Includes informal comments	• Acknowledges completion
• Conversational tone	• Formal tone
• References the invoice (when enclosed)	• May include a summary
• Identifies enclosures	• No enclosures
• Signed by the writer	• Signed by official authorized to produce
• Paper clipped to the invoice (when enclosed)	• Permanently attached behind the title page
• Loosely inserted between the title page and letter of transmittal	

The organization and style of the following sample cover letter in Figure 8.3 are businesslike. However, the tone is conversational.

Critique of the Sample Cover Letter

The sample letter in Figure 8–3 follows a standard business format in block paragraph form and has a formal opening salutation. The opening paragraph tells Mr. Marks the purpose of the letter. In the opening sentence, notice that the state, Connecticut, is spelled out because the post office abbreviation, CT, is acceptable for mailing addresses only.

The second paragraph includes unconfirmed information that the recipient would be interested in learning that would not be appropriate to include either in the report or letter of transmittal.

Mr. Franks includes a warm and courteous sentence in the third paragraph. He states that he enjoyed working with Mr. Marks and looks forward to working with him in the future.

In the fourth paragraph, instead of writing, "Please do not hesitate to call me if you have any questions," Mr. Franks wrote ,"Please call me if you have any questions." Notice the positive tone of this statement: Mr. Franks avoided using the word *not*.

The third and fourth paragraphs include items that are typically included in cover letters but not in letters of transmittal.

ENVIRONMENTAL SERVICES, INC.
123 FIELD PLACE
ASGREEN, AS 12345
(101) 234-5678

February 20, 1995

Mr. James P. Marks
Project Coordinator
ABC Design Company
Anycity, AT 23456

Dear Mr. Marks:

I have enclosed the environmental impact report of the proposed Higby Village in Yourtown, Connecticut, and a current invoice for services performed on this project.

As we discussed on the phone, the Planning Commission of Yourtown is presently considering restricting public access to shoreline property. If this is approved, the passageways from the street will not be required, and I would need to revise my report before submittal to the Planning Commission.

I enjoyed working with you on this innovative project and look forward to working with you in the future.

Please call me if you have any questions.

Sincerely yours,

Harrison Franks

Harrison Franks
President

Encl: Higby Village Environmental Impact Report
          Invoice

FIGURE 8–3
Sample Cover Letter

Mr. Franks cordially closes the letter and includes his title. *Enclosures* enumerates the supplementary items included in the transmittal. This letter is very positive and direct and avoids jargon.

INVOICE

The invoice, or statement of charges, is the statement submitted to the client to justify payment for services and supplies. Details of expenses and the time spent providing professional services permit the client to evaluate the validity of the charges. It is not common practice to enumerate separate secretarial and clerical services, because these items are usually included in the hourly billing rate for professional services. See Figure 8–4.

The recipient of the report usually authorizes payment of the invoice. However, the financial officer usually disburses the funds for payment. To expedite payment, include all applicable accounting information on the invoice, such as the account or contract number, the person to whom and the address where the payment should be sent, and your Internal Revenue Service or Tax Identification Number.

The invoice is not an integral component of the report and is attached behind the cover letter with a paper clip before insertion behind the title page.

The format of the following sample invoice appears organized and professional. And the information is easy to find.

Critique of the Sample Invoice

The letterhead clearly indicates the origin of this transmittal. The recipient identifies its contents by the heading "Invoice."

The recipient can easily locate the date of the sample invoice, client (recipient of the invoice), client's contract number, and subject of the contract.

The invoice is organized into sections and uses topical headings in bold print. "Engineering Time," because its results are included in the calculations in "Engineering Services and Expenses," immediately precedes "Engineering Services and Expenses."

The items "Total" and "Balance Due" and critical numerical values are in bold print. The items specified within each topical heading include details of functions, dates, billing rate, expenses, and retainer (a deposit or credit).

Because an invoice is a statement of charges that is the basis for payment, all hours are shown to the nearest one-tenth (e.g., 2.6) or one-quarter hour (whichever is the billing basis for the organization), and all dollar amounts are shown to the nearest one cent. The retainer is indicated with a pair of angle brackets ($<\,>$), typically used by accountants for account credits.

The invoice is easy to understand, and the numbers are multiplied and added correctly (this is a good reason for using a computerized spreadsheet to perform your calculations).

SHM ENGINEERING SERVICES
4321 WISE STREET
SMARTOWN, CO 54321
(908) 765-4321

                              INVOICE
DATE: *August 20, 1995*
CLIENT: *E & L Building Company*
CONTRACT NO.:   *654 –S*
                    2 –story building at 401 Dorothy Street

Engineering Time

Design, Jan. 17, 1995, thru June 22, 1995	*42.5 hr*
Analysis, March 12, 1995, thru July 16, 1995	*23.6 hr*
Drawing Check, June 13, 1993, thru Aug. 1, 1995	*12.3*
Total	**78.4 hr**

Engineering Services and Expenses

Engineering (78.4 hr @ $65.00/hr)	*$5096.00*
Soils Test and Report	*$ 550.00*
Total	**$5646.00**
Less Retainer	*<1200.00>*
Balance Due	**$4446.00**

Tax Identification No. 084–72–2574
Payment due in 30 days

FIGURE 8–4
Sample Invoice

The bottom of the invoice includes the tax identification number and the payment due date.

LETTER OF TRANSMITTAL

The letter of transmittal is addressed to the official who authorized the report and is signed by the official authorized to produce the report. It is the official acknowledgment of completion of the report and includes a statement of its transmittal. It mentions the title and subject of the report and can include a summary. It also can acknowledge those who assisted in preparing the report. Unlike the cover letter, the letter of transmittal is a permanent part of the report. It has a formal tone and uses a formal business letter format. It is placed behind the title page.

Organizations that routinely write reports for clients usually have a standard form or format for the letter of transmittal.

Similar to the cover letter, the organization and style of the letter of transmittal are businesslike in the sample letter of transmittal, shown in Figure 8–5. Although parts of the sample letter of transmittal are conversational, it is more formal in tone than the cover letter.

Critique of the Sample Letter of Transmittal

The sample letter in Figure 8–5 follows a standard business format but avoids the social pleasantries used in the third and fourth paragraphs of the cover letter shown in Figure 8–3.

The subject line ("Re: Model 24-C . . .") alerts the recipient concerning the subject of the report.

The first paragraph summarizes the purpose of the report and the findings. The last sentence in this paragraph adds a personal comment relating the writer's previous determination.

The second paragraph is conversational in tone with respect to future projects, "We recommend that. . . ."

Ms. Lynn cordially closes the letter and includes her title.

A copy of this letter is sent to Tara Sandor, as indicated below the closing.

Because the letter of transmittal is permanently attached to the report, there are no enclosures.

TABLE OF CONTENTS

A report longer than several pages includes a table of contents to help readers determine the subject matter of the report, its organization, and the location of sections of interest. Readers sometimes use the table of contents as an abstract (see Chapter 7).

When the major headings have subheadings and sub-subheadings, these are indented from the major and subheadings. The heading format can either use the

JONES LABORATORIES, INC.
123 Sparks Place   (305) 456–7890
ASBRIGHT AS 12345

March 12, 1995

Mr. Steven A. Whitcomb
Vice President
Alpha Electronics Company
Intown, WY 45678

Re: MODEL 24–C CONTROL PANEL FIRES REPORT

Dear Mr. Whitcomb:

Our laboratories investigated the cause of the Model 24–C control panel
fires by testing 12 prototypes under normal operational and emergency
loads. The test results included in this report consistently demonstrate inadequate wiring for emergency loads at the terminal junction. This finding
confirms our preliminary analysis.

We recommend that this junction be redesigned and retested to minimize
the possibility of future fires.

Very truly yours,

Carol Lynn

Carol Lynn, Manager
Electrical Test

Copy: Ms. Tara Sandor

FIGURE 8–5
Sample Letter of Transmittal

traditional (i.e., I., I.A., I.A.1., I.A.2., I.B. I.B.1.,) or the multiple decimal (i.e., 1.0, 1.1., 1.1.1., 1.1.2., 1.2, 1.2.1.) system. Either system is acceptable; however, most industries and government agencies prefer the multiple decimal system.

See Chapter 6 concerning the title pages of appendices and their page numbers.

In a report, all graphics except tables (e.g., sketches, charts, graphs, and photographs) are labeled *figures*. Tables are labeled *tables*.

In the table of contents, figures and tables are included in separate lists. The numbering system for each of these visual aids begins at 1, for example, Figure 1, Figure 2, Figure 3, and Table 1, Table 2, Table 3.

The headings and the numbering system in the table of contents must correspond identically to the headings and numbering system used in the body of the report.

The sample table of contents in Figure 8–6 is easy to use and is esthetically attractive.

Critique of the Sample Table of Contents

The headings are listed in the order in which the sections are presented and contain the information that identifies the topics of those sections. The list of figures and the list of tables are separate items in the sample table of contents. The appendices are labeled and listed separately from each other.

Major heading, subheading, and sub-subheadings are labeled using the multiple decimal system and are indented. (This multiple decimal system and labeled headings in the table of contents must correspond to the multiple decimal system and labeled headings used in the body of the report.) Page numbers are easy to find with a line of dots leading from the headings to the page number. All page numbers are in a right-justified column.

The sample table of contents is clearly labeled "Table of Contents" at the top of the page in large, bold letters. The titles for the list of figures and list of tables are in upper- and lowercase letters, respecting the hierarchy of the headings.

KEY CONCEPTS

- The cover materials are usually the last items to be written because they summarize the contents of reports.
- Readers may judge the quality of the contents of reports based on the cover materials. Therefore, the cover materials must positively impact readers to enhance the credibility of the report.

STUDENT ASSIGNMENT

1. Rewrite the body (headings and salutations are not included) of the following cover letter. Consider grammar, syntax, logic, structure, and content.

    In response to your letter on October 19, 1994 we are submitting the final report for the structural analysis and design of the Kiwi River Bridge.

Table of Contents

List of Figures

List of Tables

FIGURE 8–6
Sample Table of Contents

For your convenience, an invoice for the balance due for my services are enclosed. Please make payment within 14 days in accordance with our contractual agreement.

Thank you for selecting our firm, Barcal Engineering Co. I hope we will have the opportunity of providing consulting services for you again in the future.

For further questions, please don't hesitate to call me at (316) 897-1234.

2. Rewrite the body (headings and salutations are not included) of the following letter of submittal. Consider grammar, syntax, logic, structure, and content.

The reason for this Tubular Slender Columns research report is to confirm the efficiency of its use. In this report you will find an estimate of the cost-efficiency for using these columns.

It is recommended that these Tubular Slender Columns be used instead of Rectangular Slender Columns.

In Assignments 3 through 6, assume you have just written a report for submittal to a client or government agency. Write a title page, cover letter, letter of transmittal, invoice, and table of contents for one of the following (see "Types of Technical Reports" in Chapter 1):

3. A proposal to be submitted to your management or a client to develop the technology for one of the following:
   a. Minimize friction in automobile engines.
   b. Increase the strength of triangular slender columns.
   c. Reduce the effects of vibration on machine elements.
   d. Reduce the cost of electric power.
   e. Increase the stability of space vehicles.
   f. Improve methods of electrochemical machining.
   g. Update software for a computer game.
   h. Reduce aerodynamic noise.
   i. Reduce the corrosive effects of soils on pipes.
   j. Increase ship payload capacity.
   k. Improve the performance of radial tires.
   l. Increase the holding force of nails.
   m. Increase the performance of synchronous motors.
   n. Increase the strength of polyvinyl chloride.

4. A research paper to be submitted for publication to Steven Putnam of the appropriate engineering society (e.g., ASME, IEEE, etc.) on one of the following topics:
   a. Resistance of electronic circuit boards to thermal shock.
   b. Ethics in the private sector.
   c. How plasma-arc cutting works.
   d. Lubrication for high-speed vehicles.
   e. Gasification of coal as an alternative fuel source.
   f. Cost-effectiveness of high-speed mass transit.
   g. Expansion of concrete at subfreezing temperatures.
   h. Aerodynamic drag on SSTs.

    i. Strength of die casts.

    j. Legal aspects of software piracy.

    k. Speed of materials handling at airports.

    l. Applications for crystallized glass.

    m. Cost to the consumer for automotive safety.

    n. Long-range effects of toxic waste.

5. A procedural specification to be submitted to your client, Ace Engineering, for one of the following tests:

    a. Stability of Model CH-1 air-cushioned vehicle.

    b. Fire resistance of X-15 fabric.

    c. Electrical performance of a utility control panel No. 6.

    d. Motion and time study of Department 67.

    e. Efficiency of carpool lanes on Newland Freeway.

    f. Resistance of MBI Model 14 computer to power surges.

    g. Rigidity of Flisher Model 35-11 stereo casing.

    h. Performance characteristics of a GF 15 HP motor.

    i. Ultimate strength of 2024-T4 aluminum columns.

    j. Driver braking reaction distance at 70 mph.

    k. Fuel efficiency of Test Vehicle X-A.

    l. Impact resistance of Formica 80 tabletop.

    m. Shock and water resistance of Skubie 16 diving watch.

6. A feasibility study to be submitted to Michelle James, your manager, on one of the following projects:

    a. Use of heat-treated steel for front axles.

    b. Purchase of a spot welder.

    c. Relocation of Plant 3 to a distant city.

    d. Use of unsupported spans greater than 52 feet.

    e. Reduced curing time of concrete.

    f. Use of speed bumps on Phelan Street to reduce traffic.

    g. Computerized self-learning programs.

    h. Merger of the Analytical Mechanics and Design departments.

    i. Reduce production costs with an employee incentive program.

    j. Construction of a 12-unit apartment complex.

    k. Elevated temperature to increase chemical reaction rate.

    l. Use of a variable-shaped airfoil on MSB-27.

    m. Market for integrated circuits.

    n. Increased replacement time of lathes.

    o. Use of Teflon pipe to reduce friction.

CHAPTER 9

Graphics

Graphics help the reader visualize your message.

Graphics (tables, graphs, charts, drawings, diagrams, photographs) in reports present in a visually attractive and instantly comprehendible manner what the written text presents in a linear or sequential manner. Graphics make information easier for readers to understand and remember and can be used to help them interpret concepts and ideas. (See Figure 9–1.)

For many readers, graphics are the primary source of information. These readers read the written text only when they do not understand the graphics. For this reason, clear graphics to communicate your message are important.

GUIDELINES FOR USING GRAPHICS

An effective graphic presents limited information. Only one idea or concept is emphasized for a clear message. A graphic should be large enough to be read easily. However, you may use as many graphics as necessary to communicate your information.

Consider the needs of readers when selecting graphics.

- Executives use tables and charts with data and facts to communicate information to others.

- Engineers use graphs to understand relationships between variables.

- Technicians use diagrams to understand technical procedures and operations.

- Operators use flowcharts to understand sequences of events.

- Nontechnical readers use photographs, which do not require technical interpretation, to visualize objects, and charts and tables to obtain data and facts.

FIGURE 9–1
Types of Graphics

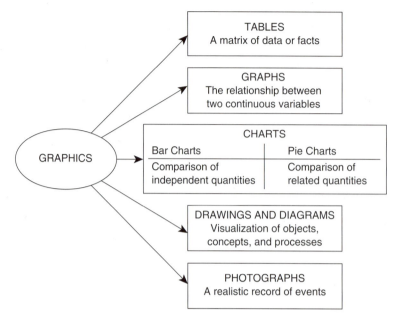

Graphics assist writers to communicate ideas and concepts. These graphics should be integrated with explanations to visually demonstrate key points in the text. For example, the following statements can be incorporated into the text:

The voltage spike seen at t = 12 seconds in Figure 4 was caused by a surge in the line current.

or

A study of the critical path diagram (see Figure 2) will help clarify the reason the analysis of the aft bulkhead was not completed by the scheduled date.

Either of these statements would require extensive description without graphics. Also, the graphics minimize the possibility of misinterpreting the message of the writer.

Please note that all graphics shown in this text are for illustrative purposes only, and the information included in these graphics is not to be used in any technical context.

Referencing Graphics

Graphics are *referenced sequentially* in the text of the report; that is, the reference to Figure 1 always precedes the reference to Figure 2. They are placed in the report in the same order they are referenced.

The sets of figure numbers and table numbers are each *numbered sequentially* beginning with 1 (e.g., Figure 1, Table 1), and usually include captions (a brief description of the contents of the graphic).

All graphics in a report are labeled figures, except for tables, which are labeled tables. However, when a report includes only several figures and tables, all graphics are usually labeled figures for the reader's convenience.

Locating Graphics

When possible, a graphic appears on the same page as its reference in the text. However, the graphic never precedes its first reference in the text. When the graphic appears on a page other than where it is referenced, the page number is included with the reference.

Reports that are only several pages long frequently place all graphics with captions at the end of the report. Also, many reports include graphics pertaining to a section at the end of that section.

Labeling Graphics

Figure numbers with captions appear below the graphics except when the figure number with caption clutters the data or scale along the horizontal axis. Then, the figure number with caption may be placed in the upper half of the graphic field. Table numbers with captions appear near the top of the graphic field. Any facts pertinent to the understanding of, and any limitations on the validity of, the information in a graphic are clearly indicated in the caption or graphic field.

Many professionals who routinely write reports develop their own graphics with computers. Most organizations have a graphic services or publications department to assist professionals in writing reports. Also, commercial copy and print shops are good sources for creating quality graphics.

Developing Graphics With Computers

Many programs are available for personal computers that make it possible for you to design professional-looking graphics to be inserted into your reports. Because of the capability, speed, and low cost of these programs, graphics can easily be generated and revised until the graphic representation is suitable for your purposes and layout.

These programs ordinarily allow you to select the desired type of graphics (e.g., table, graph), enter the data, and add relevant words, numbers, and symbols. These program capabilities (see Chapter 10) make it easy to integrate these graphics within the text of your report.

Spreadsheets (computer programs that perform repetitious calculations and print the data and results in tabular form) save time, minimize the probability of mathematical errors, and add a professional appearance to your graphics not easily attainable otherwise. Graphing programs can accurately plot the closest approximation of a smooth curve (see Step 4 in "Labeling the Scales" later in this chapter) through a set of data points, thereby eliminating the writer's estimate.

TABLES

Tables are often used because the information may be presented without interpretation (analysis or calculation) by the writer. However, numerical data are frequently interpreted by the writer and presented in the same or subsequent tables. Tables present data or facts in matrix form (a rectangular array of numerical quantities or facts) and give the reader latitude to analyze and understand their meaning. For example, data collected from a laboratory experiment are ordinarily recorded in tables (see Figure 9–2). Also, tables are used effectively for non-numerical facts (see Figure 9–3). Tables may be accompanied by graphs or charts that demonstrate the relationship of the data or facts.

Follow these guidelines to set up a table:

- Place the table number with a title or caption near the top of the table.
- Label the top of each column and the left side of each row with a title.
- When all units of measurement for the table are the same, place the units with the caption of the table (e.g., "Length, in."). Otherwise, include the units of measurement in parentheses with the column or row headings.
- For a table with numerical data, line up the decimal points in each column.
- Indicate detailed explanations for data or facts in the body of the table with a superscripted number after the data or facts in the table. The explanations appear below the table (see Figure 9–2).

FIGURE 9–2
A Table From a Student Laboratory Report

TABLE X					
Turbine Test Data and Results—Nozzle Setting = 6					
Wheel Speed, N	Brake Load	P$_{NET}$	N×P$_{NET}$	Brake Horsepower	Efficiency 3 η
(rpm)	(lb)	(lb)	(lb/min)		(%)
300	8.2	6.4	1920	0.55	52
400	7.3	5.4	2160	0.62	58
500	5.7	4.0	2000	0.57	54
600	4.5	2.8	1680	0.48	45
670[1]	3.0	1.3[2]	870	0.25	23

[1] Maximum wheel speed.
[2] P$_{NET}$ less than 2.0 is estimated.

- When tables have more than five columns and five rows, when possible, separate the data or facts into subsets to encourage reader attention. Then, either display the separation of these subsets with double or bold lines, or display these subsets in two or more tables.

Critique of the Sample Tables

Figure 9–2: Figure 9–2, a six-column by five-row table from a student laboratory report, is divided into three subsets separated by double vertical lines. The left section, "Wheel Speed," is the independent measured variable. The center section, which includes "Brake Load" and "P_{NET}," are the dependent measured variables. The right section includes three columns of calculated data (results). Although the methods for determining the calculated data are appropriately not indicated, sample calculations for determining these data should be included in a prior section of the report.

A title at the top of the table identifies the recorded data. Specific information concerning the nozzle setting is included with the title.

Each column of the table is labeled, and the units of measurement are indicated in parentheses. Each row is identified by the selected wheel speed for which the data are measured.

The decimal points for the data in each column are lined up. All data are included with the appropriate number of significant digits.

Data in the table that require additional explanations are superscripted in the body and are explained at the bottom of the table.

Figure 9–3: Figure 9–3, a non-numerical table used for determining the composition and description of different types of stainless steel, is divided into three columns. Units of measurement, not applicable for non-numerical data, are not

FIGURE 9–3

A Non-numerical Table Showing Uses of Stainless Steel

TABLE X		
Types, Composition, and Properties of Stainless Steel		
Type No.	*Composition*	*Description*
302	Basic Type Cr 18%, Ni 8%	Good formability.
304	Lower C	More weldable than 302.
316	Higher Mo	Resists salt water.
317	Higher Mo than 316	Good heat resistance, excellent corrosion resistance
405	Al added to Cr 12%	Excellent heat resistance.

included. Each row is headed with a a type of stainless steel. A title at the top of the table identifies its contents.

GRAPHS

A graph is a visual interpretation of data that shows the interrelationship between two continuous variables such as voltage and time, or stress and strain. Graphs are used for the following reasons:

- A table in a laboratory report is ordinarily accompanied by a graph that includes the experimental data points and a smooth curve approximating the path of these points to help the writer determine the relationship between the two variables studied in the laboratory. The scatter of these data points on either side of this smooth curve is a measure of the validity of the data.
- A theoretical or design graph (a smooth graph without experimental data points) helps the reader understand the relationship between the variables or is a source of technical information for design or analysis.

Use the following guidelines for drawing a graph. (Note: The use of a graphing program eliminates the need for drawing your graph. However, the responsibility for following these guidelines belongs to the writer, not the computer.)

Setting Up the Page

1. Use standard engineering graph (coordinate) paper, for example, 10 lines to the inch, for graphs. Lines spaced 1 inch apart (main divisions) should be darker than intermediate lines (divisions). This paper is available in any engineering or drafting supply store. Graphs are usually sufficiently large so that they fill at least one-half of the page. Frequently, graphs fill the entire page.

2. Plot the independent variable (the variable that is controlled during a test or selected) along the horizontal (x) axis and the dependent variable (the variable that is measured during a test or determined) along the vertical (y) axis, except when this is contrary to standard practice. Usually, the axis of the independent variable is longer than the axis for the dependent variable.

3. Allow space below the horizontal axis, and to the left of the vertical axis, for numbering the main divisions and titling the scale of each axis.

4. Darken the horizontal and vertical axes to make them prominent.

5. Place the title (figure number and caption) of the graph below the graphic field. However, when this clutters the data or scale along the horizontal axis, place the title above the top center of the graphic field.

   The caption of the graph includes a statement of its contents or its variables (e.g., "Temperature vs. Deformation"). When the caption is a statement of the graph's variables, the first variable in the statement is usually the independent variable (e.g., temperature is the independent variable in "Temperature vs. Deformation").

When the curve of only one specimen is shown in the graph, indicate the specimen's material and size (e.g., "2024-T4 Aluminum, 1/2-in. × 2-in. bar") with the caption. When the graph is titled below the graphic field, include this information with the statement of the variables. When the graph is titled in the graphic area, include this information directly below the statement of the variables.

Selecting the Scales

1. Select a scale large enough so that significant changes in the curve are apparent but small enough so that laboratory errors are not magnified.

2. Select scale units along the main divisions of the graph that are divisible by 10 for 10 lines to the inch coordinate paper and divisible by 4 for 4 lines to the inch coordinate paper so that the intermediate divisions can be readily interpolated. For example:

    For 10 lines to the inch coordinate paper, use main division scale units of 1, 2, 5, 10, or 20; the intermediate divisions are, respectively , 0.1, 0.2, 0.5, 1.0 or 2.0.

    For 4 lines to the inch coordinate paper, use main division scale units of 1, 2, 4, 10, or 20; the intermediate divisions are, respectively, 0.25, 0.5, 1.0, 2.5, or 5.0.

    This may require using less than the entire length available along the axis.

3. Number the graph 0,0 at the origin, and increase the numbers to the right, and up, respectively. When the first data point is considerably greater than zero, a section of the corresponding axis can be cut with a section-cut symbol. The plot of the curve should fill as much of the graph area as possible (also see Step 2).

Labeling the Scales

1. For the horizontal axis, place the title of the variable and the numbers so they can be read from the bottom of the graph. For the vertical axis, place the title of the variable so it can be read from the right-hand side and place the numbers so they can be read from the bottom of the graph.

    Number the axes at the main divisions of each scale only.

    Include the name of the variable measured and the units of measure, if any, in each axis title. The symbol for the variable is sometimes included. The units of measure can be placed in parentheses. For example, an axis title may read "Moment, M (ft-lb)."

2. The number of significant digits shown at the main divisions of each axis should be consistent with the precision of the measuring instruments or reliability of the results. When scientific notation is used, include the base 10 with its exponent only with the last (and therefore largest) numbers shown along the axis. Do not include the base 10 with its exponent with the title of

the axis (when you include scientific notation with the title of the axis, it is unclear whether the number read from the scale has *already been* multiplied by the power of 10 or *needs to be* multiplied by the power of 10).

3. Show data points on the graph using a 1/10-inch diameter circle with a point in the center. When multiple specimens are shown on one graph, use a different symbol with a point in the center for each specimen; for example, use a triangle or square. Use a template for drawing these symbols.

For a theoretical or a design curve drawn from points calculated by use of an equation, show the points on the graph with a point rather than a symbol.

4. When the theoretical or design curve of the engineering variables is smooth (as is typical for two continuous variables), the curve representing the laboratory specimen is also shown smooth.

The curve will probably not be able to pass through all data points; therefore, select a curve that approximates, as closely as possible, the straight segments of lines connecting the points. An equal number of points should fall above and below the completed curve. The precise method of doing this is the Method of Least Squares (see any statistics text).

The completed curve may touch the edges of the symbols but should not pass through them.

For a theoretical or design curve, draw the curve passing through all the points.

Use a French-curve to draw all continuous curves.

5. For multiple specimens shown on one graph:

Use a different line convention for each curve—solid, dash, dot-dash. Do not use a different color for each curve, because colors do not reproduce when photocopied.

Label each curve so that it reads horizontally. Use a two-section arrow to connect each label with each curve. The pointed section of the arrow should be perpendicular to the curve at the point of contact; the tail section of the arrow should be a short horizontal line drawn at mid-height from the label, and it should begin at either the front or rear of the label. A legend may be included in lieu of these labels with arrows to identify the line conventions.

See Figure 9–4 for a multiple specimen curve and Figure 9–5 for a design curve.

Critique of the Sample Graphs

Both figures are shown on rectangular coordinate graph paper; however, Figure 9–4 is shown on semilog graph paper, which may be used when the variation of the dependent variable is significantly more pronounced at the lower values of the independent variable, and these variations are to be emphasized.

FIGURE 9–4
A Multiple Specimen Curve

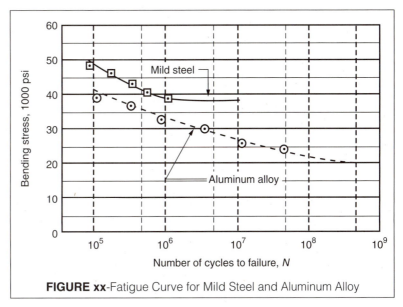

FIGURE xx-Fatigue Curve for Mild Steel and Aluminum Alloy

In the design curve of Figure 9–5, the independent variable—the variable known by the user for determining the other variable—follows standard practice and is shown along the horizontal axis. However, contrary to standard practice, in Figure 9–4, the independent variable is shown along the vertical axis. When stress (or load) is one of the variables, it is the customary practice to show it along the vertical axis. (For the purposes of this text, Figures 9–4 and 9–5 are shown smaller than one-half page.)

The horizontal and vertical axes of both figures are displayed more prominently than any division lines of the graph. The horizontal scale of Figure 9–4 does not begin at zero; however, this is inherent when using semilog graph paper, and therefore a section-cut symbol, as indicated in Step 3 of "Selecting the Scales" is not included. Also, the main divisions of each scale are numbered, and its lines are more prominent than the intermediate lines.

The scales are sufficiently large for the curves to be read and interpolated easily. In Figure 9–4, which includes experimental data points, the scale is not so large for the scatter of these points to distort an otherwise smooth curve. The main divisions of Figures 9–4 and 9–5 are divisible by 10.

The figure titles below the figures include a statement of its contents. Figure 9–5 also includes specific information concerning the applicable value of H/D.

All numbers and the titles of the variables represented by the horizontal axes can be read from the bottom of the graph. The titles of the variables represented by the vertical axes can be read from the right-hand side of the graph. The titles of all variables are descriptive and include units of measure, when applicable. The number of significant digits at the divisions of the scales is consistent with the precision of the instrumentation or reliability of the results.

In Figure 9–4, an experimental multiple specimen curve, data points are shown with points enclosed with a different symbol for each specimen, and a different line convention is used for each curve. The curves are drawn smoothly with an equal

FIGURE 9–5
A Design Curve

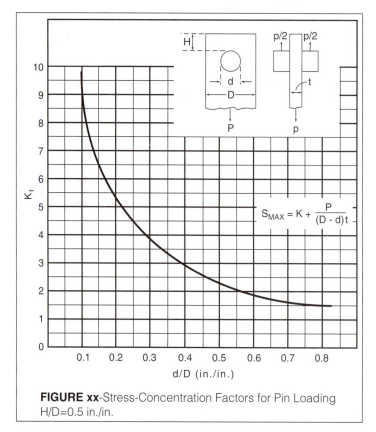

FIGURE xx-Stress-Concentration Factors for Pin Loading
H/D=0.5 in./in.

number of data points above and below each curve. These curves touch the edges of the symbols but do not pass through them. Each curve is labeled horizontally and uses a two-section arrow, which is perpendicular to the curve at the point of contact.

In Figure 9–5, a design curve without symbols, the smooth curve passes through all points. Figure 9–5 also includes a sketch demonstrating the procedure to apply the data obtained from the curve.

CHARTS

Charts are very effective for comparing quantities because they are easy for readers to understand without interpretation. There are two basic types of charts. Bar charts compare independent quantities such as time spent for design versus time spent for manufacturing. Pie charts compare dependent quantities that represent parts of a whole, such as the percentage of your inventory of parts used for each project.

Bar Charts

Bar charts show sets of horizontal or vertical parallel bars drawn to scale to compare independent quantities where each quantity represents one variation of the same set. For example, a bar chart can compare the expected lives of different automobiles for similar driving conditions.

Most writers prefer vertical (rather than horizontal) bars in bar charts, especially when the compared quantities are counted quantities, such as production of units for each model or dollars spent by each department. However, horizontal bars are used when the quantities being compared are time, such as projected time for completion of different phases of a project, or horizontal phenomena, such as stopping distance for varying road conditions.

For emphasis, the bars in a bar chart are usually shaded, cross-hatched, or otherwise marked. Bar charts sometimes show multiple sets of adjacent bars when the bars within each set are interrelated. For example, the average ambient seasonal temperatures for a period of 5 years can be shown with five sets of four adjacent bars, each bar of a set representing one of the four seasons. The marks for each bar of the set differ from the other bars in the set, but the markings in all five sets are the same. A key (a caption explaining these markings) demonstrates the designation for the markings on each of the adjacent bars.

The axis representing quantity is clearly labeled and includes the units of measurement, if any. Bars are identified by labels at their bases and, sometimes, with a key when they are dissimilarly marked. It is helpful to the reader to indicate above the bar the numerical quantity this bar represents. To avoid misinterpreting the information in the chart, bars should always begin at zero (frequently, bar charts used for advertising a product begin at other than zero to exaggerate differences).

See Figures 9–6 and 9–7 for sample bar charts.

Critique of the Sample Bar Charts

Figure 9–6, a horizontal bar chart, compares the rippability of rock by measuring the velocity of sound waves—a horizontal field measurement—through these rocks. The figure, with the number and caption at the bottom, includes a key for the dissimilar markings of each bar, which identify the ease of rippability at different wave velocities.

The vertical axis of Figure 9–6 includes the names of various types of rock to be compared. The horizontal axis is titled and includes the units of measurement. It begins at zero and is numbered on the top as well as the bottom for ease of reading. The last number along the horizontal axis includes scientific notation that clearly instructs the user to multiply all numbers by 10^3. All numbers and titles can be read from the bottom.

Figure 9–7, a vertical bar chart with multiple sets of adjacent bars, compares sets of upper, middle, and lower structural rib deformations at different impact velocities. A key is included to differentiate the different ribs at the same velocity. All numbers, the units of measurement, and the title of the horizontal axis can be read from the bottom. The vertical axis includes units and can be read from the right-hand side of the page.

Pie Charts

Pie charts show circles cut into pie-shaped wedges representing parts of a whole. Each piece of the wedge is individually shaded, cross-hatched, or otherwise

FIGURE 9-6
A Sample Bar Chart
Showing a Time-Dependent
Phenomenon

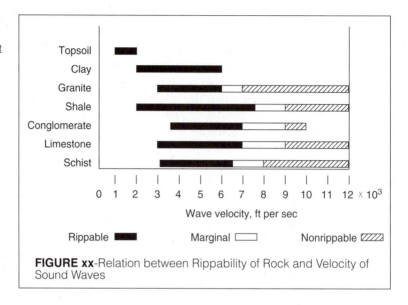

FIGURE xx-Relation between Rippability of Rock and Velocity of Sound Waves

FIGURE 9–7
A Sample Bar Chart
Showing Multiple Sets
of Related Quantities
Phenomenon

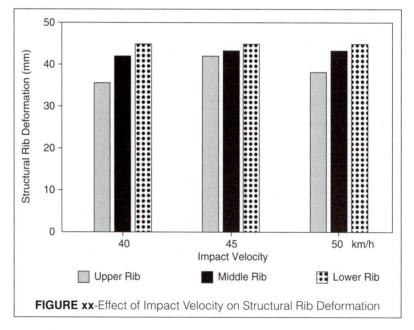

FIGURE xx-Effect of Impact Velocity on Structural Rib Deformation

marked to distinguish it from the others and can be lifted away (exploded) from the remainder of the pie for emphasis.

Each wedge is identified or labeled immediately outside the wedge and has an arrow pointing to the field. When the field is sufficiently large, the identification or label can be displayed within the field. Wedges should be arranged in sequence by size, largest to smallest, clockwise, beginning at the 12-o'clock position.

The numerical quantity and percentage of the whole represented by each wedge is shown directly below the identification. All identifications are printed horizontally. When they are placed outside the wedge, the tail of the arrow begins with a short horizontal segment at the mid-height of the beginning or end of the identification, and then continues in the radial direction approximately half-way into the wedge. The arrow terminates at a section-cut symbol. Some writers prefer to terminate the arrow at the perimeter of the wedge rather than half-way into it. See Figure 9–8 for a sample pie chart.

Critique of the Sample Pie Chart

Figure 9–8, a pie chart, compares the relative expenditures of the operation and maintenance costs for equipment for a given year. A title appears at the bottom of the figure to identify its contents.

Each wedge is identified with a title. The cost and percentage of the whole are indicated directly below the title. This information is included within the wedge when space is adequate. Otherwise, this information is placed outside the wedge and has a two-segment radial arrow terminating at a section-cut approximately half-way into the wedge.

Wedges are arranged clockwise by size with the largest wedge beginning at the 12-o'clock position. One wedge, "Tractors," is lifted away for emphasis.

Each wedge of the pie is marked differently from the others. Because each wedge is identified within the diagram, a legend is not included.

FIGURE 9–8
A Pie Chart

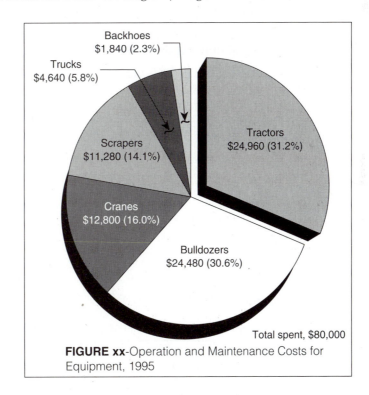

FIGURE xx-Operation and Maintenance Costs for Equipment, 1995

DRAWINGS AND DIAGRAMS

Drawings are used in reports to help the reader visualize physical objects such as pumps and buildings. They can emphasize the following:

- The interior of objects, with section views through critical planes and hidden lines
- The assembly of components, including exploded views
- Critical elements by eliminating noncritical elements
- Material, with cross-section symbols

Diagrams are also used in reports to help the reader visualize linearly dependent concepts and processes such as schematic diagrams that demonstrate how an electronic system works. (Organization charts showing management structure, and flowcharts showing computer algorithms are actually diagrams, because their elements show linear dependency.) Diagrams should be simple but contain the details that the readers need. They should be large enough to read easily.

A desktop computer can provide professional-quality drawings and diagrams at little expense.

See Figure 9–9 for a sample drawing and Figure 9–10 for a sample diagram.

Critique of the Sample Drawing and Diagram

Figure 9–9: Figure 9–9 is a pair of complementary drawings from a stress analysis. The left-hand drawing shows the shape and critical dimensions of a mechani-

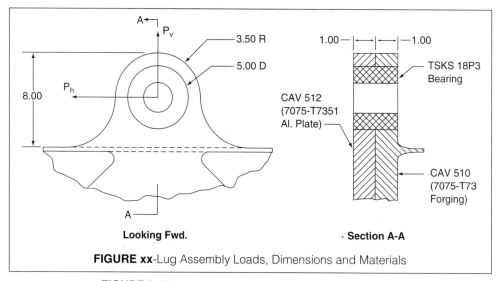

FIGURE xx-Lug Assembly Loads, Dimensions and Materials

FIGURE 9–9
A Drawing from a Stress Analysis

cal assembly and its applied loads. Section arrows (for Section A-A) tell the reader to look for a section through the plane indicated.

A section through the interior of this assembly is shown in the right-hand drawing, which includes thicknesses and drawing numbers with materials or the catalog number and part name. The different symbols in the section help the reader visualize the assembly of the components. In addition to a title at bottom of the figure, each of the two drawings is labeled to inform the reader of its perspective.

Figure 9–10: Figure 9–10, the critical path for a construction job, is a schematic diagram showing the interdependency of the various construction functions. Each box clearly represents one activity and related information.

A key instructs the reader concerning the information included in each box. The caption at the bottom of the figure informs the reader of its contents.

PHOTOGRAPHS

Photographs are excellent realistic visual aids. However, photographs are usually not mounted directly in reports for the following reasons:

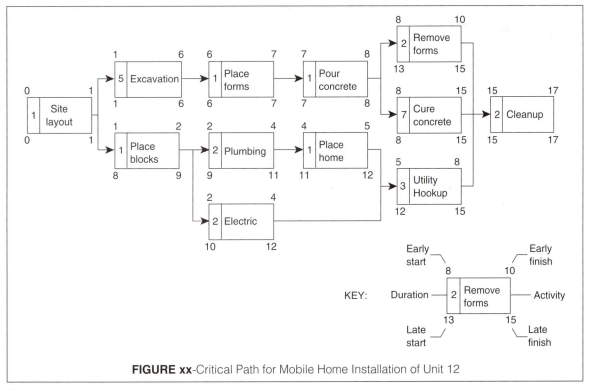

FIGURE xx-Critical Path for Mobile Home Installation of Unit 12

FIGURE 9–10
A Critical Path Diagram

- It is time-consuming to mount photographs in all copies of the report sent to the client.
- Although some rubber-based adhesives are excellent, photographs occasionally get detached and lost.
- Clients frequently photocopy the report for other readers, and many copy machines do not reproduce photographs well.

To use photographs that can be copied, do one of the following:

- Use black-and-white (preferably) photos to make a half-tone photostat to place in the original report. This photostat will reproduce on any copying machine.
- Use a computer scanner for either color or black-and-white photos to reproduce an image that may be copied on report paper.
- Use a 35-mm digital camera and import the disc image directly into your computer.

Fortunately, the equipment required for any of these processes is becoming more affordable for the serious report writer.

Emphasize important parts of photographs with labels and arrows.

See Figure 9–11 for a sample half-tone photostat and Figure 9–12 for a sample image from a computer scanner. In your report you would include a figure caption identifying the image.

Critique of the Sample Photographs

Figures 9–11 and 9–12 are both copyable on any copying machine. Figure 9–11, a half-tone photograph, includes arrows transferred from a sheet of decals for photographs purchased in a stationery store. Figure 9–12 is an image from a computer scanner. Notice that the vertical lines, which create the image, are visible and detract from the photograph's effectiveness.

KEY CONCEPTS

- Graphics add clarity to the written text without extensive description.
- Graphics should be considered as a basic mode of communication for writing reports.

STUDENT ASSIGNMENT

In the following assignments, assume you are writing a report for submittal to a client or government agency. Remember that clarity, neatness, and ease of understanding are the primary criteria for effective graphics.

FIGURE 9–11
A Half-Tone Photostat of a
Photograph (of a Spring/
Hydraulic Cylinder Assembly)

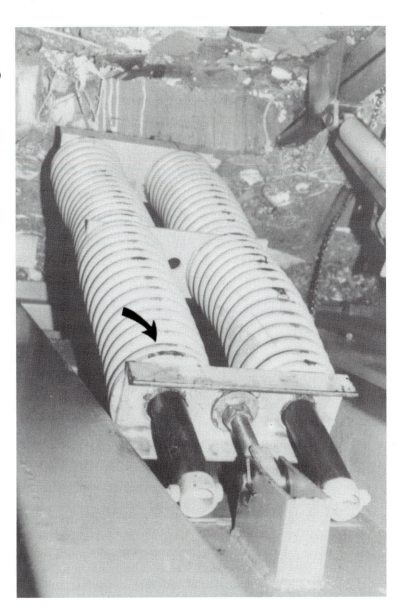

1. Prepare a table of data with at least 12 entries for a research paper to be
   submitted for publication to Steven Putnam of the appropriate engineering
   society (e.g., ASME, IEEE, etc.) on one of the following topics (reproduced
   from Student Assignment 4 in Chapter 8):
   a. Resistance of electronic circuit boards to thermal shock.
   b. Ethics in the private sector.
   c. How plasma-arc cutting works.
   d. Lubrication for high-speed vehicles.

FIGURE 9–12
An Image From a Computer
Scanner

    e. Gasification of coal as an alternative fuel source.
    f. Cost-effectiveness of high-speed mass transit.
    g. Expansion of concrete at subfreezing temperatures.
    h. Aerodynamic drag on SSTs.
    i. Strength of die casts.
    j. Legal aspects of software piracy.
    k. Speed of materials handling at airports.
    l. Applications for crystallized glass.
   m. Cost to the consumer for automotive safety.
    n. Long-range effects of toxic waste.

2. Draw a graph from test data for two specimens for a research paper in
Assignment 1. Use the full page for your graph, and show scattered data points.

Use approximately one-half of your page for Assignments 3 through 6:

3. Draw a graph to be used as a designer's technical reference for a research paper in Assignment 1.

4. Draw a bar chart showing at least three bars for a feasibility study to be submitted to Michelle James, your manager, on one of the following projects (reproduced from Student Assignment 6 in Chapter 8):
   a. Use of heat-treated steel for front axles.
   b. Purchase of a spot welder.
   c. Relocation of Plant 3 to a distant city.
   d. Use of unsupported spans greater than 52 feet.
   e. Reduced curing time of concrete.
   f. Use of speed bumps on Phelan Street to reduce traffic.
   g. Computerized self-learning programs.
   h. Merger of the Analytical Mechanics and Design departments.
   i. Reduce production costs with an employee incentive program.
   j. Construction of a 12-unit apartment complex.
   k. Elevated temperature to increase chemical reaction rate.
   l. Use of a variable-shaped airfoil on MSB-27.
   m. Market for integrated circuits.
   n. Increased replacement time of lathes.
   o. Use of Teflon pipe to reduce friction.

5. Draw a pie chart for a research paper in Assignment 1. Show at least four pie-shaped wedges in the pie.

6. Prepare a drawing or diagram of the test specimen for a procedural specification to be submitted to your client, Ace Engineering, for one of the following tests (reproduced from Student Assignment 5 in Chapter 8):
   a. Stability of Model CH-1 air-cushioned vehicle.
   b. Fire resistance of X-15 fabric.
   c. Electrical performance of a utility control panel No.6.
   d. Motion and time study of Department 67.
   e. Efficiency of carpool lanes on Newland Freeway.
   f. Resistance of MBI Model 14 computer to power surges.
   g. Rigidity of Flisher Model 35-11 stereo casing.
   h. Performance characteristics of a GF 15 HP motor.
   i. Ultimate strength of 2024-T4 aluminum columns.
   j. Driver braking reaction distance at 70 mph.
   k. Fuel efficiency of Test Vehicle X-A.
   l. Impact resistance of Formica 80 tabletop.
   m. Shock and water resistance of Skubie 16 diving watch.

7. Using a full page, prepare a set of four photographs (hollow frames without photographs are acceptable) with captions that can be used to supplement an explanation or description of one of the following:
   a. A backpack
   b. Painting a wall

c. A desk
d. A VCR
e. A subdivision
f. Setting up a tent
g. Training a dog
h. Tuning up your car
i. Dissecting a frog
j. Entering data on a spreadsheet
k. Planting a tree
l. Pouring a cement foundation
m. Operating a camcorder

CHAPTER 10

Report Visual Design and Desktop Publishing

The effect of your message can be enhanced with your visual design.

Software applications that operate on personal computers have created a flexibility to design the visual effects of your reports and to integrate graphics into their text. Today, this can be accomplished using word processing programs, whereas, until recently, this flexibility was available only with advanced desktop publishing.

Although the visual design of your report does not affect the quality of its contents, it can enhance the effect of your report on the reader. The more quickly and easily the reader can comprehend the information in your report, the greater its impact will be. You must consider the thought process of the reader who is unfamiliar with the material and unaware of any material that is presented later in the report and try to simplify this process. This can be accomplished using the principles discussed in this chapter.

The visual design should be integrated into the fabric of your final report as it is prepared and printed. Unless your organization has a standard visual design for all reports, several revisions may be required until a satisfactory design is determined. Selection of the final visual design is subject to your discretion.

Report visual design includes report and page layout, typographical format, placement of graphics, and white space.

Refer to the section entitled "Word Processors and Word Processing Programs" in Chapter 1 for a discussion of using a word processor for your report. Also, refer to "Developing Graphics With Computers" in Chapter 9, for a discussion of computer graphics including spreadsheets, and to the section entitled "Photographs" in Chapter 9, for a discussion of copying photographs into your report. Please note that the capability of computer software increases constantly, and many users have their own preferences. Therefore, it not the intent of this text to discuss any of the programs or systems presently available.

REPORT AND PAGE LAYOUT

Report pages can be either single- or double-sided. Single-sided page reports are easier for revising pages and making copies. Double-sided pages save paper and are usually used only for longer reports that require multiple copies. Most organizations have internal policies concerning the selection of single- or double-sided report pages. When bound (Figure 10–1), rather than stapled or clipped in the upper left-hand corner, these reports are easier for readers to make notes and comments (the binding provides stiffness for writing).

Report pages can be either full-page (see Figure 10–1) or two-column width (see Figure 10–2). Two-column width allows greater concentration on small segments of the information, although the narrow width disrupts continuous reading. Also, equations may not easily fit within the narrow column width. Therefore, two-column width is common only for the following reports:

- Specifications where the reader will diligently study only unfamiliar information and read all other information

- News articles where the reader can easily select to read carefully only those portions of the article that are of interest

Margins placed around the edges of the text prevent the reader's eyes from running off the page. The side margins are usually 1 inch wide. However, when the report has a binding, the inside margin is increased by as much as 1/2 inch to allow space for this binding. (See Figures 10–1 and 10–2.)

Top and bottom margins can ordinarily be set on a word processing program with headers and footers. The top margin is usually 3/4 inch, and the bottom mar-

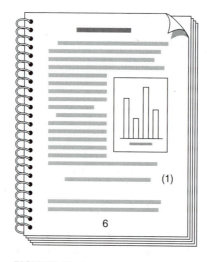

FIGURE 10–1
Full-Page-Width Report

FIGURE 10–2
Two-Column-Width Report

gin is usually a minimum of 1 inch. This may be increased to allow space for page numbers and footer notations in the bottom margin. (See Figures 10–1 and 10–2.)

TYPOGRAPHICAL FORMAT

Headings and Subheadings

Headings and subheadings provide an outline for the reader and identify the information in those sections. These headings must accurately identify the information discussed below them, using terms the reader is expected to understand. Also, these headings should use grammatically parallel structure (similar to that used in lists, see Chapter 3).

The size and style of the print of headings and use of uppercase or a combination of upper- and lowercase letters, standard or boldface letters, and underlining can establish the relative importance of these headings. To create attention, a maximum of two fonts (e.g., Helvetica, Times Roman) and three type styles (e.g., bold, italic, underline) may be used for headings.

To provide appropriate emphasis, headings are usually selected to be 2 points (1 point of print is usually 1/72 inch) larger than the letter forms of the body of the text. Although there are exceptions, it is common for the size of the headings to be 12 or 14 points, and the body of the text to be correspondingly 10 or 12 points.

These headings can either be centered on the page (see Figure 10–1) or begin at the left margin (see Figure 10–2). Their prominence should be adequate to direct the reader's eyes without overwhelming the importance of the text material included in the section below.

Body of the Text

For the body of the text of the report, upper- and lowercase letters, rather than all uppercase letters, are preferred by most readers, because their sizes and shapes are more varied and therefore more distinguishable.

Line Spacing

Line spacing, ordinarily determined by your software, is usually 2 points larger than the size of the body of the text (e.g., 12-point text will usually be 14 points between lines).

Line spacing of the text can be set at single (see Figure 10–2), one-and-one-half (see Figure 10–1), or double. Single spacing frequently has a crowded look, and unless the size of type is large or the report pages are two-column width, it is not ordinarily recommended. One-and-one-half spacing allows the reader's eyes to drop naturally at the end of each line to the beginning of the next line. Double spacing allows notes and comments to be made between the lines on the page, but the additional space between lines is excessive for comfortable reading by most readers.

Margin Justification

The margins of the text can be either left-justified (each line begins at the left margin and ends somewhere before the limit of the right margin) (see Figure 10–1) or fully justified (each line takes up the full width of the page) (see Figure 10–2). Fully justified text appears organized, but the ease of reading is decreased, because the spacing between letters may vary and all lines end uniformly at the limit line and do not provide a reference point for eye movement to the next line. Therefore, most readers prefer the left-justified text, which has an irregular right-hand margin.

GRAPHICS

Because graphics in this text are placed near their references, the reader can view and interpret the graphics as they are read. When these graphics are placed in a separate section after the text or in the appendix, the reader's attention is diverted by turning the pages back and forth, and the combined effect of the text with the graphics may be lost. Therefore, whenever possible, integrate the graphics with the text (see Figure 10–1). But remember, the first reference in the text to the graphic always precedes the graphic.

Color graphics impress the reader more than black-and-white graphics. However, when black-and-white copies are made of these graphics, the effect of the color is lost. Therefore, it may be preferable to use cross-hatching, dots, or other marking to emphasize critical areas of graphics.

Readers usually understand landscape graphics (the width of the graphic is greater than its height) more easily than portrait graphics (the height of the graphic is greater than its width), especially for those graphics that require numerical or analytical interpretation.

To minimize disrupting the continuity of reading on a page, unless a graphic uses the entire width of a page, it should always be placed adjacent to either the left- or right-hand margin rather than in the center of the page. Also, a graphic should be placed either sufficiently high on the page so that no text appears above it or sufficiently low on the page so that no text appears below it.

When graphics require an entire page and are rotated 90° in a report, they are always rotated counter clockwise so the bottom of the graphic is on the right-hand side of the page.

WHITE SPACE

White space is any part of a page that is blank and used to separate ideas. Included in the white space are the spaces between paragraphs, sections, and within the page.

For block paragraphs (i.e., when the beginning of each new paragraph begins at the margin), an additional space should be included between paragraphs (see

Figure 10–2). For indented paragraphs (see Figure 10–1), this additional space is not necessary but may be desirable.

To allow the reader to transition the thought process from the previous section, a triple (for a new heading) or double (for a new subheading) (see Figure 10–2) space should always precede a new section. A heading and subheading should be followed by a smaller space before the beginning of the text of that section (see Figure 10–2). When each section is several pages in length, it is common to begin each new section at the top of a new page (see Figure 10–1).

A widow is a new heading or subheading followed by two lines or less of text at the bottom of a page. An orphan is two lines or less of text at the top of a page followed by a new heading or subheading. Both widows and orphans create a discontinuity in the reader's thought process and should be eliminated. When a widow occurs

- The right margin of the page with the widow may be decreased sufficiently to move the widow to the top of the following page.

- Or a page break can be inserted at the new heading or subheading (see Figure 10–3).

When an orphan occurs:

- The right margin of the page immediately before the orphan may be increased sufficiently to move the orphan to the bottom of that page (see Figure 10–4).

- Or a page break can be inserted at the beginning of the last paragraph of the page immediately before the orphan to move that paragraph to the page including the orphan.

Equations and quotations included with the text should be preceded and followed by a double space, and left- and right-hand margins should be increased to

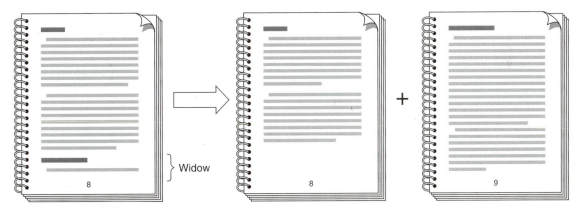

FIGURE 10–3
A Widow and How to Avoid It

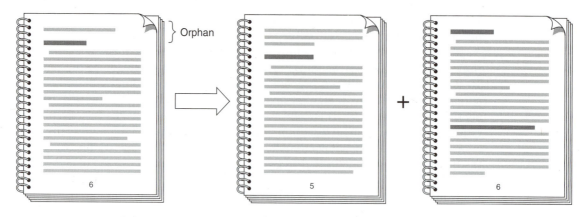

FIGURE 10–4
An Orphan and How to Avoid It

emphasize their importance (see Figure 10–1). Also, graphics that are included with the text should be surrounded by a border of white space (see Figure 10–1).

KEY CONCEPTS

- The impact of your report can be enhanced by a visual design that helps readers comprehend its information.
- The intended use of your report ordinarily will determine the report and page layout.
- Headings and subheadings are the outline of your report and direct the reader through its contents.
- Graphics, when integrated with the text, have greater impact than when placed in a separate section at the end of the report.
- White space creates a pause in the reading and therefore is used effectively to separate ideas.

STUDENT ASSIGNMENT

For each technical report used for the Student Assignment in Chapter 1:

1. Review and discuss the effectiveness of the report and page layout.
2. Review and discuss the effectiveness of the typographical format.
3. Review and discuss the effectiveness of the graphics.
4. Review and discuss the effectiveness of the use of white space.
5. If you were designing this report, would you do it differently?

Applications for Students and Professionals

The first two sections of this text introduced the principles of technical writing and described the elements of technical reports and documents. Section III builds on these fundamentals by demonstrating how to write student and professional reports.

WRITING YOUR REPORT

Each chapter of Section III recommends sections with headings for the type of report discussed. These headings are guidelines for beginning reports. Occasionally, new sections are introduced that were not specifically discussed in Section II. When this occurs, the wording of the heading clearly indicates the content of the material discussed in that section.

A preliminary draft of headings provides a framework for writing the sections of the report. The purpose of this draft is to reveal to the writer the organization of the report and to identify the discussions contained within these headings. Therefore, these headings are modified when they do not properly identify the contents of the discussions. The headings in the student samples therefore may differ from the headings recommended in the chapter.

This text is not intended to provide you with a fill-in-the-blanks format for technical writing. To the contrary, you must carefully study each assignment so that you can select the most appropriate style, format, and sections for your report.

The sections recommended in the discussions in Section III are at a level commensurate with student abilities and experiences. Therefore, the suggested sections may be fewer in number but broader in scope and include fewer details than the sections used in professional reports.

The first chapter in Section III discusses the components of student laboratory reports. A sample laboratory report with a discussion is included. Subsequent chapters of Section III discuss the purpose, audience, structure, and content of different types of professional reports. Annotated student responses to class assignments are included.

METHOD OF INSTRUCTION IN SECTION III

The critiques and annotations to the sample responses demonstrate the principles discussed in the text.

It is important for you to recognize that the samples included in Sections III and IV are typical responses for new writers and may contain deficiencies. To deter you from making these same deficiencies in your writing, it is intended for you to study these typical responses with their critiques and annotations and learn from them. Therefore, rather than using these samples as models, you should carefully study each sample with the critique and annotations before responding to the assignment given by the instructor.

Section III discusses four categories of writing:

- Student reports are discussed in Chapter 11 ("Student Laboratory Reports").

- General business writing is discussed in Chapters 12 ("Business Letters and Memos"), 13 ("Periodic, Progress, and Trip Reports"), and 14 ("Personnel Performance and Recommendations for Raises").
- Planning reports are discussed in Chapters 15 ("Bids and Proposals"), 16 ("Specifications"), 17 ("Activity and Product Evaluations"), 18 ("Feasibility Reports"), and 19 ("Environmental Impact Statements and Reports").
- Analytical reports are discussed in Chapters 20 ("Technical and Sociotechnical Articles"), 21 ("Experimental and Test Laboratory Reports"), 22 ("Scientific and Engineering Analyses"), and 23 ("Research Reports").

Student Laboratory Reports

Laboratories teach the scientific method.

In a scientific laboratory, the validity of a hypothesis can be tested from a given set of facts or data. This is known as the scientific method; beginning with the known, the unknown can be discovered.

Because the scientific method plays a significant role in educating engineering and science students, this chapter discusses the reports that result from students' experiences in the laboratory.

Many of the elements of technical reports discussed in Chapters 5 through 9 apply to laboratory reports. This chapter discusses the following specific sections of laboratory reports:

- Description of the test item
- Equipment and apparatus
- Procedure and process descriptions
- Data and calculations

It also includes a sample student laboratory report so that students may understand what laboratory instructors generally expect.

CONTENTS OF LABORATORY REPORTS

Laboratory report requirements vary from instructor to instructor, but a formal laboratory report may contain the following elements shown in the order in which they typically appear (see Figure 11–1). Figure 11–2 presents a complete student laboratory report that includes each of these elements:

Title Page

Table of Contents

FIGURE 11–1
Elements of a Student Laboratory Report

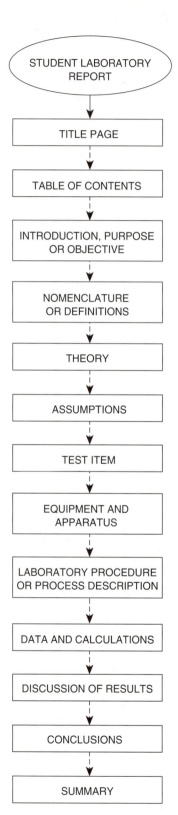

STUDENT LABORATORY
REPORT

TITLE PAGE

TABLE OF CONTENTS

INTRODUCTION, PURPOSE
OR OBJECTIVE

NOMENCLATURE
OR DEFINITIONS

THEORY

ASSUMPTIONS

TEST ITEM

EQUIPMENT AND
APPARATUS

LABORATORY PROCEDURE
OR PROCESS DESCRIPTION

DATA AND CALCULATIONS

DISCUSSION OF RESULTS

CONCLUSIONS

SUMMARY

Introduction, Purpose, or Objective

Nomenclature or Definitions

Theory

Assumptions

Test Item

Equipment and Apparatus

Laboratory Procedure or Process Description

Data and Calculations

Discussion of Results

Conclusions

Summary

DESCRIPTION OF THE TEST ITEM

A brief physical description of the tested specimen, equipment, structure, or system confirms the identity of the test item to readers.

4.0 Soil Samples

Soil samples were obtained from Spadra 323 borrow pit. The unit weight is 110 lb per cu ft in situ.

EQUIPMENT AND APPARATUS

When the laboratory equipment is commonly used by laboratories, or consists of simple components such as strain gauges and thermometers, you may include, without descriptions, a list of equipment. This is typical for student laboratory reports. Otherwise, when special equipment with which readers may not be familiar is used for the experiment, it is customary to describe the equipment and apparatus to help these readers understand the procedure and the test data.

Equipment

Rockwell Hardness Tester

The Rockwell hardness tester is a manually operated test machine used to determine the surface hardness of metal specimens. The load is applied to the specimen by an indenter through a system of levers and a counterweight. A dashpot incorporated in the loading system insures gradual loading. A dial gauge is calibrated to read a hardness number, which is inversely proportional to the depth of indentation; that is, a greater indentation will give a lower hardness number. B, C, and F scales on the dial gauge are each used for materials of different ranges of hardness.

LABORATORY PROCEDURE OR PROCESS DESCRIPTION

Similar to instructions, laboratory procedures use the imperative voice and include a step-by-step format with step numbers. Alternatively, process descriptions are the technical descriptions of laboratory procedures and may be stated in the present or past tense.

The choice of including either a laboratory procedure or a process description depends on the instructor's preference. Both are discussed here and shown in Figure 11-2 on pp. 147-164.

Laboratory Procedure

In a laboratory procedure, the sequence of events determines the sequence of the statements.

Example

Confusing: Press the STOP button when the load drops abruptly. [This may prompt the operator to press the STOP button before there is an abrupt drop in the load.]

Clear: When the load drops abruptly, press the STOP button.

A sample procedure follows.

How to find the cosecant of an angle on an HP-11C calculator.

1. Press the <ON> button.
2. Press <g> and <7> to put the calculator in the degree mode.
3. Press the digits that represent the angle. Press <ENTER>.
4. Press <SIN>.
5. Press <1/x>.
6. Read the cosecant in the display.

Critique of the Sample Procedure

Each instruction begins with an imperative verb. Step 2 has an explanation to help the reader understand a potentially confusing instruction; otherwise, explanations are not given. The names of the buttons are clearly identified in angle brackets. The last step tells the reader to read the cosecant, whose value is the objective of the procedure, in the display.

In this sample procedure, inappropriately completing one of the sequential steps would necessarily affect the results of all subsequent steps, and the final result.

Process Description

The process description is either an explanation or history of the procedure. The present tense is used for an explanation, and, alternatively, the past tense is used for a history. The following sample process description is an explanation; alternatively, a history is shown in brackets.

> *The process described below is [was] used to determine the cosecant of an angle using an HP-11C calculator.*
>
> *After the calculator is [was] turned on and put in the degree mode, the angle is [was] entered into the display from the keyboard. The sine (the reciprocal of cosecant) of the angle is [was] then displayed. The reciprocal of the value in the display is [was] then determined and read from the display.*

Critique of the Sample Process Description

The opening statement tells the reader the purpose of the process. The second paragraph discusses the method used to obtain the desired result.

The second paragraph begins with a chronological transition, *after.* Chronological transitions occur twice more in this paragraph with the word *then*.

Because the reader may not understand the purpose of finding the sine of the angle, the writer explains it. Notice that the explanation of sine is explained with respect to the cosecant (the reciprocal), rather than the functional reason for performing that operation (namely, a cosecant button is not available on the calculator).

DATA AND CALCULATIONS

The data are the numerical quantities determined from reading the instruments during the experiment. These are usually recorded on data sheets—tables with the nomenclature of the controlled and measured parameters of the experiment, respectively, as the row and column headings.

The calculations are the analysis of these data and are the numerical quantities used for understanding the results of the experiment. Because of the interrelationship between the experimental data and calculations, the experimental data and calculations are usually shown in the same tables. Frequently, spreadsheets (see Chapter 9) are used for this purpose.

The data and calculations must be presented in an easy-to-follow format, with explanations, to facilitate review by the readers. A bold or double line clearly separates the test data from the calculations.

Ordinarily, the parameters of the experiment are dependent, and graphs illustrate the relationship of this dependence. The guidelines on visuals included in Chapter 9 discuss the presentation of tables and graphs.

Some instructors recommend that you place the original data sheets either before the calculations or in the appendix of the report and that you copy the experimental data into tables that include the calculations.

SAMPLE STUDENT LABORATORY REPORT

Figure 11–2, at the end of the chapter, presents a complete student laboratory report illustrating the items that may appear in such a report. The following discusses each element.

Although the sample student laboratory report in Figure 11–2 does not conform to all the standards of presentation discussed in this text, the presentation is clearly understood.

Presentation

This report has a professional appearance. It is organized so that information is easily found; it has a title page and headings for the different sections and is neatly presented.

The report was written with a computer except for the curves on the graphs, which were drawn using a drafting program. Using the latest computer technology for writing a report enhances the credibility of the writer.

The headings are printed in bold print for emphasis.

The Cover Material

Cover material consists of the cover page and the table of contents.

- Most colleges and universities recommend that students use a standard laboratory report folder instead of a cover page. The cover page of this report uses the recommended format for reports. However, it is not clear whether the date specified on the cover is the date that the report was submitted, or the date that the laboratory experiment was performed. Also, it is common practice to include the names of the members of Group 2.

- Each of the headings of sections in the table of contents should be introduced with a section number (e.g.,"1. Introduction") for ease of identification. These section numbers are then used to head the sections in the body of the report.

- The series of dots leading across the page in this table of contents helps the reader to select the correct page number.

- Page numbers help the instructor locate the material included in the report.

The Beginning of the Report

The beginning of the report comprises the introduction, the objectives, and definitions.

- The introduction discusses traditional arch forms and explains the advantage of the semicircular arch, the subject of this report.

- Each component of the objective is preceded by a bullet to emphasize its importance. The objective of this laboratory experiment is to help the reader understand a physical concept rather than to demonstrate compliance with a specification. This is typical of a student laboratory experiment.

- The Definitions section provides the reader with specific terms and their meanings. This list is located near the beginning of the report for the convenience of the reader.

The Body of the Report

The body of the report includes the theory, test specimen and setup, a list of test equipment, the test procedure, and all data and calculations.

- The theory section includes the development of the equations with a figure that clearly identifies the meaning of all symbols. This figure includes a number and a caption. Equations are identified with equation numbers.

- The test specimen and setup are discussed in the same section because the specimen forms an arch, which is an integral part of the test setup.

  The discussion of the test setup is included to familiarize the reader with the equipment. A figure is included to demonstrate the placement of the equipment for the setup. This discussion is ordinarily not included in student laboratory reports when the readers are expected to be familiar with the equipment.

- A list of test equipment, always included in laboratory reports, is for the reader who is interested in duplicating the laboratory experiment.

- The test procedure is presented in two formats (A and B). Either format may be acceptable for student laboratory reports. Your instructor will indicate which format to use.

  Format A is a list of instructions in the imperative that the reader can use to duplicate the laboratory experiment. The details of the operation are included.

  Format A is divided into four parts, with each part having a component leading to the ultimate objective. The numbering system starts with Part I and continues through Part IV because the laboratory experiment is conducted in sequential order. If the parts of the laboratory experiment were not necessarily sequential, a separate numbering system would be used for each part.

  Format B is a historical process description. It emphasizes what was performed, rather than the method for performing the operation. If Format B had been written in the present rather than the past tense, Format B would be an explanatory process description.

Because Format B requires an understanding of the procedure, most instructors prefer this format. However, professional reports almost always include Format A because it is practical.

- Data and calculations are shown in the same tables. The data can be differentiated from the calculations by the nomenclature of the column headings.

- All but one of the graphs include theoretical as well as experimental values. The theoretical values are indicated with solid squares, and the theoretical curves should necessarily pass through all solid squares.

  The experimental values are indicated with hollow squares. The curve through the experimental values should be the best approximation of smooth curves and therefore do not necessarily pass through all hollow squares. Also, the experimental curves do not cross the outline of any square.

  More than one curve is drawn on each graph. Each curve has its own graphical representation (e.g., solid line or dashed line).

The Ending of the Report

The ending of the report consists of the discussion and conclusion and the summary.

- Because the conclusions are partially based on the performance of the test, the discussion and conclusions are included together. This section compares the experimental results with the theory. It discusses an irregularity with the operation of the test. It also explains the reason the discrepancy in values in Part IV may have occurred.

- The summary discusses the validity of elastic analysis for arch-type bridges.

TWO-HINGED ARCH

STRUCTURAL MATERIALS LABORATORY

CE 306

State Polytechnic University

May 30, 1995   Presented to: Prof. R. Ertel

Group 2   Submitted by: Hovel Babikian

FIGURE 11–2
Sample Student Laboratory Report (pp 147-164)

Table of Contents

FIGURE 11–2 (*continued*)

Introduction

The three traditional forms of an arch are semicircular, parabolic, and semielliptical. Generally, the choice of which should be used is determined by the separation distance and the difference in the elevations of the terrain surrounding the bridge. For example, a semicircular arch is often found spanning a large ravine. The large rise-to-span ratio of this arch reduces the required horizontal restraint of the structural system.

Objectives

The objective of this laboratory test is to compare experimental values of the horizontal reactions of a semicircular arch with theoretical values to determine if

- The arch behaves as a linear elastic structure
- The principle of superposition is valid for several simultaneous loads

Also, the following mechanical behavior is investigated:

- The horizontal reaction that is due to a point load moving across the arch
- The influence line for the horizontal reaction by analysis of a model

Definitions

Pinned Support: A structural support that prevents movement in the horizontal and vertical directions but does not prevent rotation about the hinge.

Roller Support: A structural support that offers resistance to movement in the direction perpendicular to the supporting surface beneath the roller. There is no resistance to rotation about the roller or to movement parallel to the supporting surface.

1

Theory

The horizontal reactions at the pinned supports of a semicircular arch are determined by the following equation:

$$H = 2/\pi\{V_a \int_o^{\psi a} (\sin \theta_a \cos \theta_a)\, d\theta_a + V_b \int_o^{\psi b} (\sin \theta_b - \sin \theta_b \cos \theta_b)\, d\theta_b\}$$

(See Figure 1 for definitions of symbols)

For a single point load, the equation simplifies to:

$$H = 4Wn(1-n)/\pi \tag{1}$$

For a uniformly distributed load of $\underline{w}$ per unit of horizontal length, the pin reaction is:

$$H = 4wR/3\pi \tag{2}$$

The theory of superposition can be used to determine $\underline{H}$ for combined loads.

2

FIGURE 11–2 (*continued*)

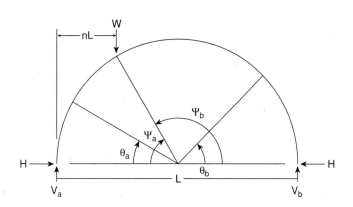

Where:

H = Horizontal thrust (N)
L = Span of the arch
V_a = Vertical reation at the left hinge
V_b = Vertical reaction at the right hinge
W = Applied load
θ_a = Angle measured to the first stirrup on the left
θ_b = Angle measured to the first stirrup on the right
Ψ_a = Angle measured to the stirrup where the load is applied
Ψ_b = $180 - \Psi_a$
n = $\frac{1}{2} - (\cos \Psi_a)/2$

Figure 1 - Definition of Symbols

Test Specimen and Setup

The test specimen is an 8-mm $\times$ 40-mm mild steel bar formed into a semicircular arch of 0.5-m radius. One hinged end is mounted on ball-bearing rollers that slide on a horizontal ground plate that supports only a vertical reaction. The other end is hinged by a pin, which supports horizontal and vertical reactions. Seven load hanger stirrups are attached to the arch at intervals of 0.125-m (horizontally) across the arch. A dial gauge indicates the outward displacement of the hinged end when a load is applied to the arch. A horizon-

3

tal thrust is applied by adding loads to a hanger attached by a cord to the hinge (see Figure 2). The applied loads to the hanger make the resultant horizontal thrust $\underline{H} = 0$.

Test Equipment

The semicircular arch (see Figure 2) includes
1. HST.501   Bracket assembly
2. HST.502   Track plate assembly with dial gauge
3. HST.503   Reaction load hanger
4. HST.504   Cable assembly
5. HST.505   Load hangers
6. HST.5b   Load stirrup
7. HAC.5m   Dial gauge assembly

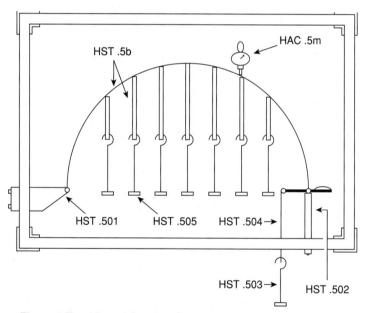

Figure 2-Two-Hinged Semicircular Arch Showing the Test Setup

4

FIGURE 11–2 (*continued*)

Test Procedure

[**Note:**   The test procedure can be shown in the report as a list of instructions (Format A) or as a process description (Format B). Both methods
have been included. Consult your instructor for which method to use.]

Format A

Part I

1. With a load hanger on any one of the stirrups near the middle of the
   arch, note the dial gauge datum reading at the roller hinge.
2. Apply a vertical load to the hanger causing the datum reading to
   change (see Figure 3).
3. Apply a load to the horizontal reaction hanger to restore the dial
   gauge reading to the datum value, creating zero displacement at the
   hinges.
4. Record the horizontal reaction required to restore the dial gauge
   reading to the datum value.
5. In five equal increments, reduce the hanger load to zero.
6. Plot the horizontal reactions versus the dial gauge readings on graph
   paper.

Part II

7. With all the load hangers in position, apply a point load to any stirrup.
8. Determine the horizontal thrust required to return the dial gauge
   reading to the datum.
9. Remove the applied point load and apply a point load of a different
   magnitude to another stirrup.
10. Record the horizontal thrust required to return the dial gauge read-
    ing to the datum.
11. Simultaneously place the two point loads on their respective stirrups.
12. Record the horizontal thrust required to return the dial gauge read-
    ing back to the datum.

5

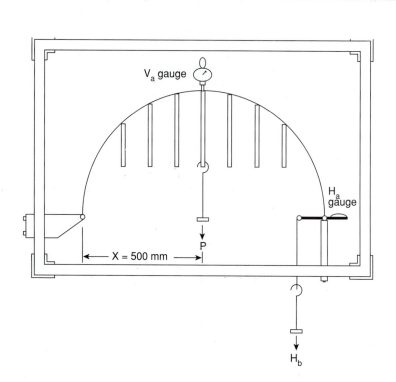

Figure 3-Two-Hinged Semicircular Arch Showing Applied Loads

Part III

13. With all load hangers in position, apply a load of 50 Newtons to the left-most stirrup.

14. Record the horizontal thrust needed to return the dial gauge reading to the datum.

15. Remove the 50-Newton load.

16. Record the horizontal thrust needed to return the dial reading to the datum as a 50-Newton load is applied on each stirrup respectively from left to right.

6

FIGURE 11–2 (*continued*)

17. Plot the results as an influence line for the reaction (i.e., plot the horizontal thrust along the y-axis and the load position along the x-axis).

18. Apply 20 Newtons to each of three adjacent hangers simultaneously.

19. Measure the horizontal thrust for the load of 20 Newtons as described in Part I.

Part IV

20. Record the dial gauge on a bracket to measure the vertical displacement of the platform at the left-most loading stirrup.

21. Record the datum dial reading at the bracket, and the horizontal reaction dial gauge.

22. Apply a load of 50 Newtons to the horizontal thrust load hanger.

23. Record the readings of both dial gauges.

24. Repeat the procedure for each loading stirrup position from left to right.

25. Plot the ratio of vertical to horizontal displacement along the y-axis and the position of the vertical displacement dial gauge along the x-axis an influence line for the horizontal thrust.

Format B

This procedure was used to determine the thrust and displacement characteristic for a 50-Newton moving load across a two-hinged arch span.

The test apparatus was calibrated by placing a vertical hanger load near the middle of the arch and measuring the horizontal thrust required to restore the horizontal displacement to zero. The vertical hanger load was then incrementally reduced to obtain a plot of vertical hanger load versus horizontal thrust.

Two different vertical hanger loads were independently applied to any two vertical hangers to determine, for each hanger load, the horizontal thrust required to restore the displacement to zero. To determine the cumulative horizontal thrust required to restore the horizontal displacement to zero, the same two vertical loads were then simultaneously placed on the same two hangers.

To draw an influence line of the horizontal thrust for a 50-Newton load moving across the span, a 50-Newton load was simultaneously placed on each vertical hanger to determine the horizontal thrust.

7

A 20-Newton load was placed on three adjacent vertical hangers to simulate a partial uniform load. The horizontal thrust was then measured.

To draw an influence line of the ratio of vertical to horizontal displacement for a 50-Newton load moving across the span, a 50-Newton load was placed on each horizontal thrust load hanger to record the vertical and horizontal displacements at the left loading stirrup.

Data and Calculations

The data and calculations are presented in Tables 1 through 3 and Figures 4 through 7.

TABLE 1   Comparison of Theoretical and Experimental Horizontal Loads

X (mm)	P (N)	Theoret Hb (N)	Experim. Hb (N)	Difference (%)
500	10	3.1	3.3	3.6
500	20	6.37	6.7	5
500	30	9.55	10	4.5
500	40	12.73	13.2	3.5
500	50	15.92	16.5	3.5

8

FIGURE 11–2 (*continued*)

TABLE 2 Varying Loads at Successive Stirrups

X (mm)	P (N)	Theoretical Hb (N)	Experimental Hb (N)	Difference (%)
125	10	1.39	1.45	4.1
250	15	3.58	3.65	1.9
375	20	5.97	6	0.5
500	25	7.96	8.15	2.3
625	30	8.95	9.15	2.2
750	35	8.35	8.45	1.2
875	40	5.57	5.65	1.4
		SUM = 41.77	SUM = 42.5	1.7
125	40	5.57	5.65	1.4
250	30	7.16	7.3	1.9
375	20	5.97	6.15	2.9
500	10	3.18	3.3	3.6
625	20	5.97	6.15	2.9
750	30	7.16	7.3	1.9
875	40	5.57	5.65	1.4
		SUM = 40.58	SUM = 41.5	2.2

9

TABLE 3   Same Load at Alternating Stirrups

X (mm)	P (N)	Theoret Hb (N)	Experim. Hb (N)	Differ- ence (%)	Ha Gauge (mm)	Va Gauge (mm)	Va/Ha (x.001)
125	50	6.96	7.2	3.3	0	0	0
250	50	11.94	12.4	3.7	10.7	0.021	2
375	50	14.92	15.2	1.8	10.85	0.077	7.1
500	50	15.92	16.2	1.7	11	0.1425	12.9
625	50	14.92	15.2	1.8	11	0.073	6.6
750	50	11.94	12.1	1.3	11.3	0.023	2
875	50	6.96	7.1	1.97	0	0	0

10

FIGURE 11–2   (*continued*)

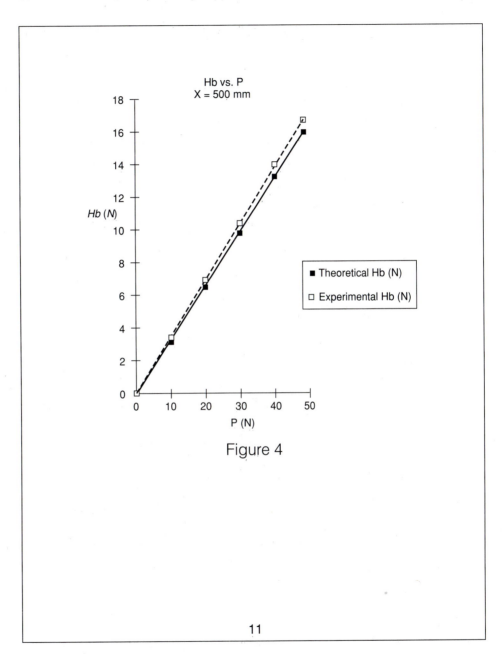

Figure 4

11

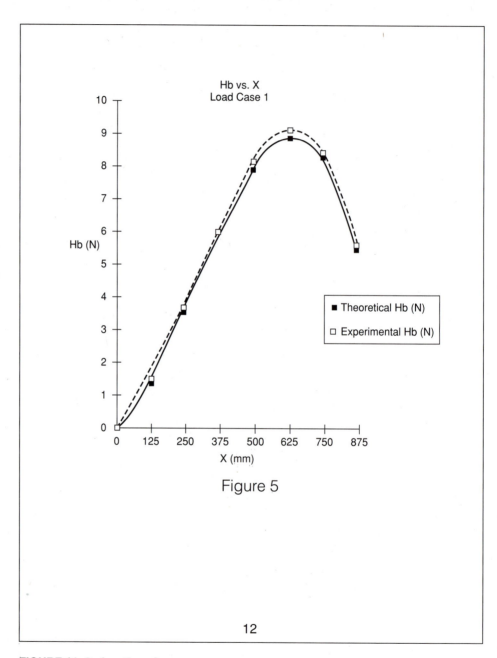

Figure 5

12

FIGURE 11–2 (*continued*)

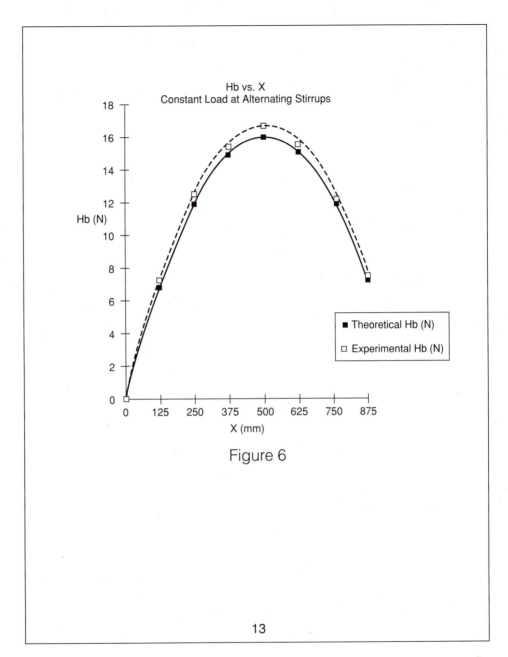

Figure 6

13

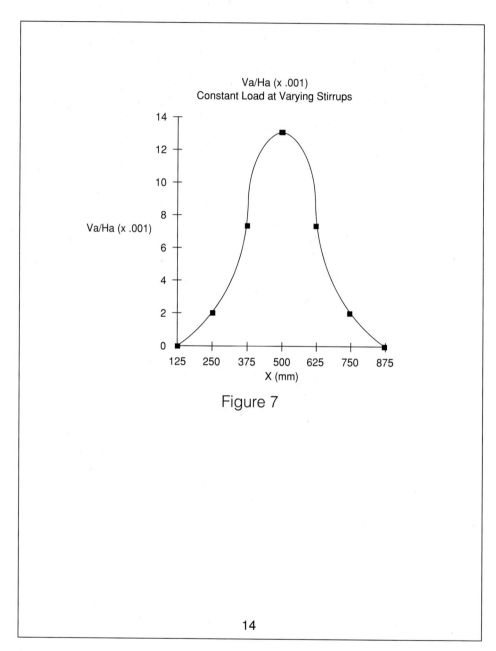

Figure 7

14

FIGURE 11–2 (*continued*)

Discussion and Conclusion

Part I and II

It is observed from the experimental curves in Part I that the semicircular arch behaved elastically. Figure 4 shows that each time that the load was increased by 10 Newtons, the experimental horizontal reaction increased linearly by approximately 3.3 Newtons. Figure 4 also shows that there is a maximum difference of 4.5% between the theoretical and experimental result, which is acceptable. The theoretical values were obtained using Equation 1.

In Part II, the data from the several loading conditions also demonstrated elastic behavior of arches. Table 2 shows that identical loads placed alternately at symmetric hangers produced identical horizontal reactions. The horizontal reactions at the two supports are identical when the load is placed in the middle hanger of the apparatus. As the load moves toward either support, the magnitude of the vertical reaction closer to the load will be greater. This is due to the increased resistance the support provides to maintain structural static equilibrium.

Parts III and IV

In Part III, by placing the 50-Newton load at alternative hangers, it is demonstrated that the horizontal reaction influence line is symmetrical about the influence ordinate corresponding to the load placed at the middle stirrup. This is consistent with the behavior of structures that was shown in Parts I and II. The relative error in Part III was a maximum of 3.7%, which is acceptable.

In Part IV, the experimental results did not agree with the theory. The ratio of the vertical displacement to horizontal displacement should yield a value that is close to the horizontal resistance force that the support exerts on the structure. The discrepancy between the theoretical and experimental displacements is probably due to the test apparatus. Often, dial gauges do not compensate for frictional forces and give incorrect readings when friction is not negligible.

Summary

The conclusions stated in Parts I through IV indicate that the two-hinged semicircular arch apparatus validated the analysis for the behavior of arch-type bridges. The semicircular bridge demonstrated the elastic characteristics of an arch-type bridge, and it also supported the superposition principle.

16

FIGURE 11–2 (*continued*)

KEY CONCEPTS

- Laboratory experiments are the primary basis for validating scientific hypotheses.
- Understanding the scientific method is a key component in the education of engineers and scientists.

STUDENT ASSIGNMENT

1. Rewrite the following description of the test item. Review grammar, syntax, logic, structure, and content.

   The A36, 16 in. $\times$ 4 in. $\times$ 1/4 in. steel plate specimens will be used for testing. One specimen will be preheated to 150°F. One specimen will be at room temperature. One specimen will be cooled at 40°F.

2. Rewrite the following equipment and apparatus section. Review grammar, syntax, logic, structure, and content.

   The torsion machine is of the mechanical gear type. It can be operated manually or by an electric motor. The load is transmitted from the moving head to the stationary head, which, in turn, transmits it to the pendulum. The swing of the pendulum balances the applied torque and, at the same time, actuates a scaled lever calibrated to read the torque applied.

3. Rewrite the following procedure. Review grammar, syntax, logic, structure, and content.
   1. Fit the specimen to be tested into spherical end bearings, then measure the unsupported length of the column for center to center of the bearing balls. Measure diameter of the specimen.
   2. Balance the scale beam with the procedure indicated on page 14 of its operation manual.
   3. Set up the specimen in the testing machine vertically. Adjust the screw sockets to ensure free rotation of the ball bearings. Apply load using the lowest speed of .05 inches per minute.
   4. While the load is being applied, the balance lever must be continuously moved outward so as to balance the load. Catch the reading when there is a sudden drop of the scale beam and a simultaneous bowing of the specimen.

4. Rewrite the following process description as an explanation. Review grammar, syntax, logic, structure, and content.

The centrifugal pump operates at four different constant speeds: approximately 1200, 1400, 1600, and 1800 rpm. At each of these speeds the flow will be varied by means of the discharge valve. Five rates of discharge, in equal increments varying from zero to the maximum, are tested at each of these speeds.

For each discharge rate, the data is measured to compute total head, input horsepower, and output horsepower.

5. Write a description of the test item for one of the following:
   a. A structural aluminum member
   b. An automobile engine or transmission
   c. The human heart
   d. An extension ladder
   e. An electronic circuit board
   f. A suspension bridge
   g. A conveyor belt
   h. A backhoe, front-loader, or crane
   i. A wing of a small airplane
   j. A theodolite
   k. Seawater
   l. An artificial limb
   m. An electric drill

6. Write an equipment and apparatus section for one of the following:
   a. A manometer
   b. A thermometer
   c. A micrometer
   d. A stethoscope
   e. A Wheatstone bridge
   f. A balance
   g. A Galvanometer
   h. A gate, globe, or ball valve
   i. A tachometer
   j. A tension scale

7. Write a procedure for one of the following:
   a. Rotating a television aerial on your roof for best reception
   b. Adjusting lead in a mechanical pencil for drawing
   c. Fitting a backpack for comfort
   d. Inflating the air pressure in your tires to 30 psi
   e. Adjusting the right-hand side mirror of your car without remote control
   f. Increasing the attitude of a drafting table from the flat position
   g. Regulating tap water for comfort

8. Write a process description as an explanation or a history for one of the procedures in Assignment 7.

CHAPTER 12

Business Letters and Memos

*Letters and memos communicate timely information intended
for a limited audience.*

Writing business letters and memos is routine for all professionals. Because these correspondences are personal and may be the only communication with the recipients, they can create lasting impressions of the writers.

PURPOSE AND AUDIENCE

Business letters and memos convey information. They differ from reports; their information is timely and addressed to a limited audience, and they usually require actions and responses from the recipients. The writers communicate information that can be direct, as when the information is for the readers' use; indirect, as when the information requires action; or persuasive, as when the writer hopes that the information will be acted on.

These communications can be addressed to clients, vendors, supervisors, subordinates, associates and colleagues, and so on.

Letters Compared With Memos

Although letters and memos are formatted differently, they serve the same purposes. Business practice dictates under which circumstances each correspondence is used. Sometimes, memos are less formal in tone than letters.

Correspondence sent to a recipient outside the company should be in the format of a letter. Correspondence sent via interoffice mail is ordinarily in the format of a memo; however, formal interoffice correspondence (e.g., transfer of a contractual obligation) can also be in the format of a letter.

CONTENT

A letter or memo generally consists of three elements:

1. An introductory paragraph that presents the subject and purpose of the correspondence and the desired outcome, if any. It establishes a personal (use of *I*, *you*, *we*, etc.) tone.

2. The information to be conveyed, including all explanations and details the reader may need to understand this information.

3. A closing statement that explains to the reader what action you desire or what action you will take; information concerning whom to contact and how (e.g., include telephone number and working hours) for more details; and an acknowledgment of the attention the reader has given to the subject.

TONE

Be Positive

Consider the benefits of your message to readers before you begin to write your correspondence. When possible, convey these benefits in the opening paragraph.

The tone of a correspondence may vary depending on the results hoped to be achieved and the audience. For example, when the subject of the correspondence is a matter under dispute with a client, use a firm tone to state your position and the action you intend to pursue. However, when the subject of the correspondence is a request, use a friendly and courteous tone to indicate your appreciation of the recipient's response.

To create a positive attitude, always emphasize what can, will, or needs to be done, even when your message is negative (e.g., the denying of a request for funds or the reporting of an unexpected result in a laboratory test).

Negative words (e.g., *no*, *not*, *never*, *will not*, *cannot*, and *but*) and words with negative prefixes (e.g., *mis-*, *misunderstood*, *mistake*; *in-*, *ineligible*, *inept*; *un-*, *unlikely*, *unfit*; *dis-*, *disapprove*, *discourage*; and *im-*, *impossible*, *impatient*) cause readers to become defensive and resist understanding your perspective. Instead of using these words, discuss the proposed action.

Examples:

Negative: Thus far, we have *not* complied with the velocity requirements. Therefore, we are requesting an extension of thirty days.

Positive: We request an extension of 30 days to enable us to comply with the velocity requirements.

Negative: I *discourage* the approval of the drawing submitted by our supplier.

Positive: Please request our supplier to revise the drawing for compliance with our acceptance criteria.

Judgmental words (e.g., *abandon, abrupt, biased, careless, cheap, complaint, deny, evade, exaggerate, failure, low, neglect, negligence, only, pointless, problem, regret, ruin, senseless, should, sorry, useless, vague, waste, weak, worry,* and *wrong*) may require readers to interpret the consequences. Rather, discuss specifics or the proposed action.

Example

Judgmental: The structural failure of the bracket was due to the *cheap* quality of the steel.

Action oriented: The bracket will be redesigned using an alloy steel to prevent structural failures.

Use the Active Voice

Consider the following sentences:

Passive: The pulley assembly will be analyzed immediately.

Active: I will analyze the pulley assembly immediately.

Both sentences convey the same basic information. However, because the actor is not identified in the first sentence (passive), this sentence is somewhat vague and lacks energy. Because the actor is identified in the second sentence (active), this sentence is to the point, easy to understand, and displays enthusiasm.

Frequently, professionals inappropriately write in the passive voice because it sounds more intellectual, it avoids the use of personal pronouns such as *I* or *we*, and it avoids accountability.

The active voice clarifies the actor. Use the active voice in business letters and memos unless the passive voice is justified by any of the following circumstances:

- The actor has not been identified. For example, "The nuclear power plant was sabotaged."

- Revealing the actor is discrediting. For example, "A mistake was made."

- The identity of the actor is unimportant or irrelevant. For example, "The Chief Engineer was promoted to Vice President."

FORMAT

The block and semiblock formats and the parts of letters are shown in Figures 12–1 and 12–2, respectively. Either format is acceptable for business communications, but most organizations have a preferred format. The format and parts of memos are shown in Figure 12–3.

325 Otto Road   Heading
Atlanta, GA 52240
October 3, 1995   Date Line

Doublespace Twice

Ms. Judy Priddo   Inside Address
Personnel Manager
Bingo Corporation
325 Mountain Drive
Honolulu, HI 90211
Doublespace
Subject: Your letter of September 18, 1995   Subject Line
Doublespace
Dear Ms. Priddo: Salutation
Doublespace

Doublespace

Doublespace

Doublespace
Sincerely yours,   Complimentary Close

Doublespace Twice

(Mrs.) Meg Cook   Signature Block
Doublespace
Enclosure: Test report
Doublespace
Copy: Ms. Janet Yoshioka

FIGURE 12–1
Block Letter

325 Otto Road
Atlanta, GA 52240
October 3, 1995

Doublespace Twice

Mr. James Good, Manager
Bingo Corporation
325 Mountain Drive
Honolulu, HI 90211
Doublespace
Subject: Your letter of October 20, 1995
Doublespace
Dear Mr. Good:
Doublespace

Doublespace

Doublespace

Doublespace
Sincerely yours,

Doublespace Twice

Amanda J. Black
Doublespace

Copy: Ms. Francis Canars
      Mr. Edward Schwartz

FIGURE 12–2
Semiblock Letter

SALUTATIONS

In formal correspondences such as in most letters, regardless of your personal familiarity, address the recipients with their titles or professional designations and surnames such as

Dear Mr. _____:

Dear Dr. _____:

Dear Prof. _____:

Unless you are certain of the gender of the recipient, replace subtle sexist salutations such as "Dear Drafts*man*" with nonsexist salutations such as "Dear Drafter." Also, in business correspondence, "Dear Miss _____" or "Dear Mrs. _____" is inappropriate, because these salutations presume the marital status of the recipient. Rather, use "Dear Ms. _____."

When the surnames of the recipients are unknown, try to determine them. This will help your correspondence to receive the most attention. Otherwise, use functional titles such as "Dear Vice President of Engineering," "Dear Researcher," and "Dear Homeowner." Do not use vague salutations such as "To Whom It May Concern," because they do not capture the recipients' attention.

In informal correspondences such as in most memos, you may address the recipients by their given names (e.g., "Dear Dave").

Use a colon, rather than a comma, after the salutation for business communications.

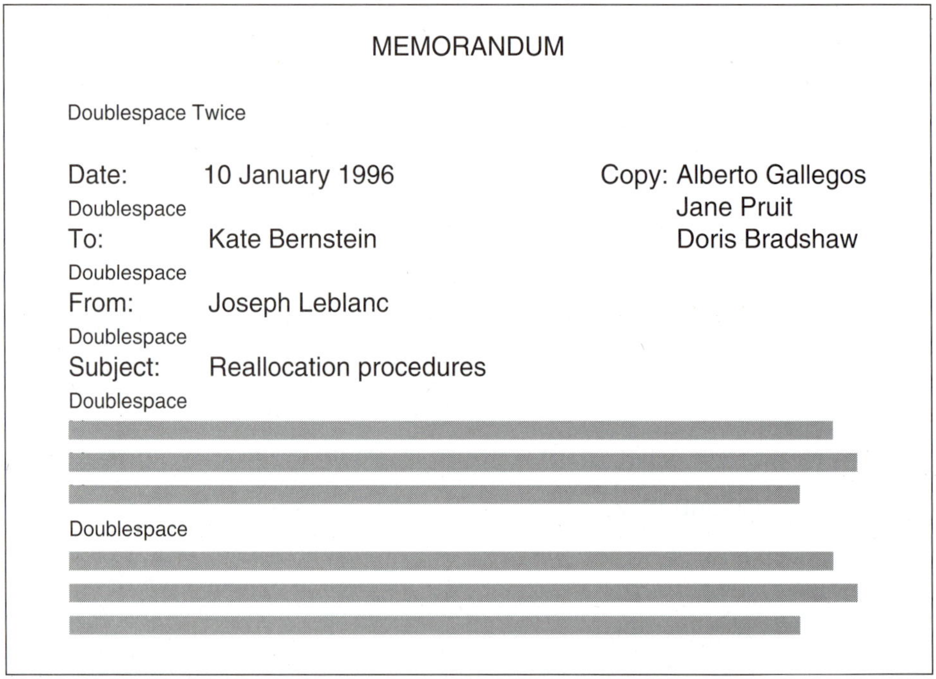

MEMORANDUM

Doublespace Twice

Date:	10 January 1996	Copy: Alberto Gallegos
Doublespace		Jane Pruit
To:	Kate Bernstein	Doris Bradshaw
Doublespace		
From:	Joseph Leblanc	
Doublespace		
Subject:	Reallocation procedures	
Doublespace		

▓▓▓▓▓▓▓▓▓▓▓▓▓▓▓▓▓▓▓▓▓▓▓▓▓▓▓▓▓▓▓▓▓▓▓▓▓▓
▓▓▓▓▓▓▓▓▓▓▓▓▓▓▓▓▓▓▓▓▓▓▓▓▓▓▓▓▓▓▓▓▓▓▓▓▓▓
▓▓▓▓▓▓▓▓▓▓▓▓▓▓▓▓▓▓▓▓▓▓▓▓▓▓▓▓▓▓▓▓▓▓▓

Doublespace

▓▓▓▓▓▓▓▓▓▓▓▓▓▓▓▓▓▓▓▓▓▓▓▓▓▓▓▓▓▓▓▓▓▓▓▓▓▓
▓▓▓▓▓▓▓▓▓▓▓▓▓▓▓▓▓▓▓▓▓▓▓▓▓▓▓▓▓▓▓▓▓▓▓▓▓▓
▓▓▓▓▓▓▓▓▓▓▓▓▓▓▓▓▓▓▓▓▓▓▓▓▓▓▓▓▓▓▓▓▓▓▓

FIGURE 12–3
Memorandum Format

CLOSINGS

In letters, include a complimentary closing such as "Very truly yours" or "Sincerely yours." Follow this closing with your signature and your name typed below it. In memos, sign your initials or name adjacent to your printed name in the heading.

In letters, following your typed name, when applicable, include a list of enclosures preceded by "Encl." or "Enclosure(s)" and a list of others who will receive a copy of the correspondence preceded by "Copy." In memos, when applicable, include the list of enclosures below your message and a list of others who will receive a copy in the heading.

TYPES OF BUSINESS CORRESPONDENCE

The most common types of business letters and memos that you will be expected to write as a professional include the following:

Requests for Information

The recipients of requests for information (e.g., a request for material properties from a research library or current regulations from a government agency) frequently have no obligation to respond. Therefore, you must convince them of the importance of this information to you.

Begin your correspondence by clearly stating that you request information and the subject of this information. Explain your purpose for obtaining this information and why you selected the recipient as the source.

Carefully explain the requested information to help the recipient understand your needs and to simplify the response. Use a list of numbered questions when you seek several bits of information. Offer to supply the recipient with the results of your use of this information. To promote goodwill for your organization, follow up this offer by supplying this information.

End your correspondence by telling the recipient where the requested information should be sent and with an appreciative closing statement such as "I will appreciate this information."

Requests for Action

Requests to support an activity or program with time or money (e.g., a request to become active in a professional society or to subscribe to a magazine) are similar to requests for information, except that you are asking for a greater commitment from the recipient, and therefore you must be more persuasive.

Arouse recipient interest in the opening paragraph with a statement or question concerning an existing need (e.g., to be a well-informed professional) rather than creating a new need (e.g., to attend a conference). Address the immediate or long-term benefits that may be derived from this commitment.

Keep the opening paragraph brief to encourage the recipient to begin reading. Include sufficient details to generate interest; however, do not discourage interest by including unnecessary details.

Close your correspondence with a clear statement of the immediate action the recipient should take to support this activity or program. Emphasize the importance of a prompt response and facilitate this response by enclosing a reply card or an envelope. Alternatively, state that you will contact them by a specific date.

Responses to Requests for Information or Action

Courteous responses to requests for information or action foster positive images of your organization to your readers, especially when you have no obligation to respond to these requests. Some requests for information or action require responses, such as invitations to activities that limit participation and claims that ask for action resulting from delivery of an unsatisfactory product or performance.

Begin your correspondence by stating that you are responding to the reader's request. Then, provide the requested information, or indicate the action that you will take. Close the correspondence by expressing concern for having complied with the request of the reader with a statement such as "Please call (or write) me if you need further information."

When the request cannot be complied with, explain the reason, and when possible, offer an alternative to the reader for obtaining the requested information or action.

Following is a sample of an opening paragraph from a letter requesting action:

Dear Borrower:

In the event that you have not received your payment coupon book in time to mail your first payment, please send this form with your check or money order to:

Thaler Savings and Loan
123 State Street
Yourtown, MM 80888

We are pleased you have chosen Thaler Savings and Loan. . . .

Critique of Sample Opening Paragraph of a Letter Requesting Action

The writer clearly bypasses the purpose of this letter in this paragraph by abruptly instructing the reader how to compensate for the possible inefficiency of its system rather than informing the reader that a request for a loan has been approved. The tone of this correspondence could easily alienate the borrower in a long-term financial relationship.

The generic salutation implies that the borrower is merely a file without an identity in spite of a significant financial commitment. The borrower's name should always be included in any business transaction.

The opening paragraph is in the imperative tone and, except for the word *please*, demonstrates a cold and negative attitude (i.e., the use of the word *not* in the first line) toward the borrower. Also, the opening phrase, *in the event*, is wordy and should be replaced with *if*. A more appropriate text would be the following:

Dear Ms. Weinberg:

Thank you for doing business with Thaler Savings and Loan.

Since the recent approval of your loan by this office, we notified our Milton City office to mail your payment coupon book. You should receive this book very shortly.

If you do not receive your payment coupon book in time for us to receive your first payment by October 10, you may send your payment with this form to the following address:

Thaler Savings and Loan
123 State Street
Yourtown, MM 80888

We are pleased you have chosen Thaler Savings and Loan. . . .

This letter is courteous, informs the recipient of positive news, and explains clearly, but warmly, the action required by the recipient if a subsequent correspondence is delayed.

This correspondence was an opportunity for the lender to foster a positive working relationship, which the sample failed to accomplish. Rather, the sample tended to create an impersonal, defensive relationship with the borrower.

Claims

Claims correspondences (e.g., claims of dissatisfaction with performance of a product or of a product defect) should courteously attempt to persuade the recipients to understand your point of view and the actions you desire as the result of an unsatisfactory product or performance. Use a friendly tone of voice, and assume the recipients are reasonable and fair persons. Be firm in your requests, but demonstrate a pleasant attitude to obtain the desired outcomes.

Before stating your claim, begin the correspondence with a background of the facts that led to the present situation. State the details of your problem and the actions that you have taken to eliminate or alleviate your dissatisfaction. Close your correspondence with a suggested course of action for the recipient.

Professional Invitations

Because invitations (e.g., to be a guest speaker, participate in a panel discussion, or attend a meeting) ask the recipients to expend time and effort to benefit your organization, you must be tactful in presenting your requests.

Begin your correspondence with the nature of the invitation, the reason for selecting the recipient, and the benefit to your organization of participating in this event.

Include the date, time, and location of this event, arrangements that you will make (e.g., hotel reservations), and financial considerations (e.g., reimbursement of travel expenses or an honorarium, which is a gratuity to show appreciation for professional services).

Close the correspondence by expressing hope that the recipient will accept the invitation and by encouraging a timely response to expedite planning of the event.

Instructions

Frequently, the recipients of instruction-type correspondences (e.g., product recalls or revision of company procedures) have no prior obligation to follow the instructions, and therefore you must explain their importance. A conversational tone will encourage cooperation.

In the opening paragraph, discuss either your desire to achieve a common goal (as between departments or organizations working on a joint project) or your concern for the satisfaction and success of the reader as a user (as in the instructions for a more efficient or safer use of a product). Explain the expected results and benefits to the reader that can be achieved by following the instructions.

See Chapter 5 for writing instructions. To encourage compliance with the steps in the instructions, include brief explanations when their purpose or importance is not obvious to the reader. However, clearly separate the steps from the explanations. This can be accomplished by using italics, parentheses, or another literary mechanical device for the explanations.

Express your hope for the expected results and benefits, and extend an offer of additional assistance in the closing paragraph.

Form Letters and Memos

Form letters and memos are used to reach a wider audience than other correspondences. Because a form letter or memo can be one of the many types discussed previously, use the applicable principles as indicated.

Unfortunately, form correspondence is read less seriously than personalized correspondence because of its lack of a targetted audience. Because a memo is work related and reaches a narrower audience than a letter, a form memo is frequently reviewed for content and filed for future reference, whereas a form letter is usually disposed of immediately unless there is a specific reason for the reader to retain it.

To maximize the attention of your reader, care must be given to the format (e.g., using bold headings and indented paragraphs). Use good-quality stationery and first-class mail for letters. When possible, use a computer program to include inside addresses and salutations on each correspondence and individually addressed envelopes (rather than mailing labels); use a personal signature in blue ink (to distinguish it from the black print of the copy) below the complimentary closing.

When personalized addresses and salutations are not feasible, use the narrowest salutation possible to address your audience (e.g., use "Dear Microbiologist" rather than "Dear Scientist"). When a generic salutation is used, the inside address is usually not included.

Cover Letters and Memos

Cover letters and memos introduce a report to the intended readers. See the section "Cover Letter or Memo" in Chapter 8 for a discussion.

E-Mail

E-mail (electronic mail) transmitted via the Internet (an electronic communication network) is sent by a computer to the computer of the recipient. It is less formal than a letter or memo but more formal than a telephone call. It usually follows the format of a memo. Because it can be sent any time of the day without interrupting the recipient, it is convenient, especially when the sender and the recipient are in different time zones. However, because others may have the ability to retrieve personal messages, only nonconfidential information should be communicated by e-mail. Also, e-mail may be recoverable by others after deletion.

SAMPLE ASSIGNMENT—MEMO

Write a memo to your supervisor to request approval to attend a professional seminar in Hawaii. You may create purposes and contexts for any additional information you may need.

See Figures 12–4 and 12–5.

Critique of Sample Memo 1

The writer in Sample 1 (see Figure 12–4) has difficulty explaining the purpose of the request. The subject, "Seminar in Hawaii," implies that the location rather than the theme of the seminar needs approval. The opening sentence requests the money to attend this seminar. The theme of the seminar is not mentioned until the third sentence.

The second paragraph appropriately discusses the potential benefits to the company by attending this seminar. But suddenly, the discussion reverts back to the cost of this attendance.

This sample has the fatal flaw of not requesting any specific action by the reader or a date for the response.

Critique of Sample Memo 2

Sample Memo 2 shown in Figure 12–5 is well written. The subject in the heading makes the purpose of the memo clear. The opening paragraph addresses the subject, dates, place, and presenting organization. The second paragraph addresses the potential benefits to the company.

The cost issue, addressed in the third paragraph, is appropriately treated as a detail, not the subject, of the memo. The attention of the reader is thereby given to the benefits of the seminar rather than the cost.

Memorandum

Date: October 18, 1996

To: Purchasing Department

From: Mukhtiar Singh   *MS*

Subject: Seminar in Hawaii

I request $900.00 to attend a professional seminar in Hawaii on the 16th through the 18th of November. The seminar will discuss the corrosion of steel and methods to prevent corrosion and to protect against weathering.

This seminar will benefit our company because the design of bridges is our specialty, and corrosion is a problem. The estimated cost of the trip includes airfare, hotel accommodations for three nights, meals, and seminar registration fees.

This type of memo should be addressed to the writer's supervisor, not the purchasing department.

The subject is stated incorrectly. The theme of the seminar is more important than the location and therefore should be the subject.

FIGURE 12-4
Sample Memo 1

Memorandum

DATE: January 12, 1995

TO: Steven Morris
Design Supervisor

FROM: Martin Rodriguez *MR*

SUBJECT: Approval to attend an expert systems seminar

I request approval to attend a seminar on computerized expert systems on March 20 and 21 in Hookoo, Hawaii. This seminar will be presented by the American Society of Civil Engineers as part of their extended education program.

A general overview of expert systems for civil engineers will be presented the first day of the seminar. Specific applications will be discussed the following day. I hope to use the knowledge that I gain at this seminar to increase the capability of our department to design complex structural systems.

The cost of attending this seminar is estimated to be $1230. This includes $600 for airfare, $250 for registration, $225 for hotel accommodations, $75 for meals, and $80 for car rental.

Because enrollment is limited, applications must be submitted by February 28, 1995. Please notify me of the status of this request by February 14, 1995.

FIGURE 12–5
Sample Memo 2

The closing paragraph tells the reader what action is needed and gives the reader a critical date for response.

SAMPLE LETTERS

See Figures 8–2 and 8–3 for a sample cover letter and the accompanying text in Chapter 8 for a critique, and see Figure 8–5 for a sample letter of transmittal.

KEY CONCEPTS

- Letters and memos are routine communications for professionals. Their information is timely and addressed to limited audiences.
- To the reader, letters and memos are the most personalized form of communication. They provide an excellent opportunity to foster goodwill.
- Effective letters and memos may be your only opportunity to create lasting impressions on your readers.

STUDENT ASSIGNMENT

Write a letter or memo (as appropriate) for one of the following. You may create purposes and contexts for any additional information you may need.

1. A request to the accounting department of your organization for an advance of expenses for a site inspection in a distant city.
2. A notification to employees of a change in overtime policy.
3. A denial of a request from a supervisor in another department for you to participate in a joint project.
4. A loan of a 20-page report to an associate.
5. A claim to your company's supplier concerning the malfunctioning of a mass-produced audio component, assembled into an entertainment center and distributed by your company.
6. A request to colleagues urging political action on a political-technical issue.
7. A denial of a request for a refund by a dissatisfied purchaser of your widget intended for recreation that performs as promised.
8. An invitation to a young professional to join your professional organization.
9. A response to a client for instructions for operating a system supplied by your company.
10. A cover letter to a client for an enclosed test report.
11. A request for payment from a client on a 90-day past due invoice.

12. An invitation to a professor at your alma mater to be a guest speaker at a professional meeting.

13. A request for an opinion from a supervisor in a different department concerning the validity of a method of analysis for your application.

14. A claim to a supplier concerning the inability of its "maximus hardwarus" to perform as well as promised.

15. A request for technical information from the research department of a high-tech company.

16. A notice and instructions for a recall to all purchasers of your company's widget that is due to a design defect.

CHAPTER 13

Periodic, Progress, and Trip Reports

Management and clients stay informed with periodic, progress, and trip reports.

Professionals are responsible for communicating their accomplishments and setbacks with management and clients. They communicate this type of information with periodic, progress, and trip reports.

PURPOSE AND AUDIENCE

Periodic reports summarize the tasks accomplished on all projects since the last regular reporting period. Immediate supervisors and managers frequently require periodic reports once a month or at other regular intervals. These reports emphasize completed tasks rather than anticipated completions. Because periodic reports are intraorganizational, they are ordinarily in memo format.

Progress reports summarize the tasks accomplished on one project only. They address completed tasks, projected completions, and may include recommendations for current and future phases of a project. Progress reports are written either as memos or letters depending on the intended audience; ordinarily, memos are written for intraorganizational personnel, and business letters are written for clients. Frequently, these reports partially fulfill contractual requirements. Progress reports may be required either at regular intervals or at predetermined milestones of project completion.

Trip reports are progress reports that discuss the events and accomplishments of business trips. Most organizations use a standard format for these reports. When this is not the case, follow the guidelines in this chapter for progress reports. Expense reports, separate from trip reports, are prepared for accounting purposes and ordinarily are on a preprinted form that is completed by the professional.

Writers can use periodic, progress, and trip reports to review recent accomplishments and to determine the direction of their effort during the next reporting period. This often renders quality projects that are completed on time and at the intended cost.

Project reports (not discussed in this chapter) summarize the events of projects after these projects are completed. They are written as either memos or letters depending on the intended audience. Because the projects are completed, these reports are historical accounts. See Chapter 20, "Technical and Sociotechnical Articles," for information about reporting completed events.

The terminology used to identify periodic and progress reports varies from organization to organization. Some organizations refer to periodic reports as project status or progress reports; to progress reports as project or status reports or project updates; and to project reports as project summaries. When given the task of writing one of these reports, always clarify the needs of the recipient to determine the content of your report.

CONTENT

Periodic and progress reports enable you to communicate your accomplishments and problems since the most recent report. To maintain your credibility, it is important to be honest in reporting immediate and potential problems. Because you have greater knowledge of this project than anyone else, it is appropriate to recommend alternatives for resolving these problems. On the contrary, trip reports emphasize the outcomes of recent events but may include potential future actions.

Periodic Reports

Periodic reports document the current status of all your projects so that supervisors and managers can coordinate the work of the professionals in their groups and departments. These reports emphasize what has been accomplished since the last reporting period. They may also include the anticipated progress during the next reporting period.

Periodic reports are frequently less than one page long so that they can be read quickly. Progress on any one project is usually treated in only one or two paragraphs. For easy identification, headings are used for each project.

Progress Reports

Progress reports keep supervisors, managers, and clients abreast of the accomplishments on their projects. These reports discuss details concerning dates for completion, compliance with the quality requirements of the product or project, and deviation from budgeted cost. Although the reader is probably familiar with your project, do not assume prior knowledge from previous progress reports when writing these reports.

Example

Wrong: I received the parts that we ordered on schedule.

Correct: I received the six 3/4-inch nozzles from Hazzle Fluid Products on February 24, the scheduled delivery date.

Include anticipated progress during the next reporting period in progress reports. A Gantt chart (a horizontal bar chart of activities versus time, see Figure 13–1) is frequently used to accomplish this. When progress is behind schedule, or problems are anticipated, these problems are addressed. Alternative courses of action and recommendations to expedite completion may be suggested. Although progress reports are written at designated times and emphasize progress during the current reporting period, they also address progress since commencement of the project when this information is necessary to understand the current progress.

The sections of progress reports include the various activities required to complete the product or project (e.g., preliminary layout, design, analysis, fabrication of prototype).

Sample Analysis Section of Seepee Bridge Progress Report

A computer stress analysis has been performed of the bridge and was completed. The analysis includes the limitations that are due to buckling. The method of the analysis was confirmed by hand calculations.

Critique of the Sample Analysis Section of a Progress Report

The method of expression in the first sentence, "A computer stress analysis has been performed of the bridge and was completed," is redundant by including both *performed* and *completed*. Obviously, when the stress analysis is completed, it was performed. The phrase "has been performed" should be eliminated, and the sentence could more appropriately read "We completed a computer stress analysis of the bridge."

The information in this sample needs to be more definitive. The student does not indicate which computer program was used for the analysis. Was it SAP 90? EASE 2? The writer assumes the reader either knows the program or is not concerned about it. In either case, it is good practice to remind the reader of the program.

The mention of the limitations that are due to buckling is ambiguous. Is this a limitation on the engineering properties of the bridge or on the capability of the computer program to account for buckling?

The meaning of "The method of analysis was confirmed by hand calculations" is vague. The reader might think that the analysis was performed twice, first with the computer program and then with hand calculations. The student may have meant "Using the Method of Joints, a hand analysis, we verified the validity of the computer program for this application by randomly selecting and calculating the stresses of four members of the bridge at different locations."

FIGURE 13–1

Gantt Chart for a Construction Project. The black portions of the bars indicate completed work.

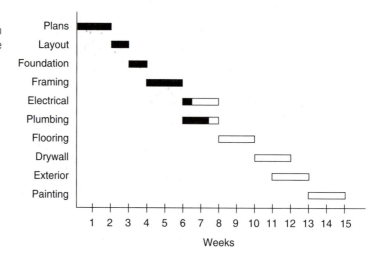

Because the analysis was completed, there is no need to address anticipated work on this phase of the project. However, the date of completion may be important to the reader and should be included.

Consider the following revision of the above sample of analysis section:

> We completed the analysis of the structural members of Seepee Bridge on January 30 using SAP 90, a finite element stress analysis program. The analysis assumes that the ends of all members are pinned and includes the load limitations of the structural members that are due to the effects of buckling.
>
> By using the Method of Joints, we verified the validity of the computer program for this application by randomly selecting four members of the bridge at different locations and calculating their stresses.

Trip Reports

Trip reports inform others in your organization or your client of the events and accomplishments of a business trip. These reports include dates, places, purpose, events attended, accomplishments, and decisions for desired or required future actions that result from this trip.

Trip reports are very similar to progress reports except that, frequently, additional potential benefits of a trip do not become apparent until the trip is in progress. Therefore, these reports may also include the desired or required future actions that have resulted from these trips.

Sample Assignment—Periodic Report

Write a periodic report to your instructor concerning your accomplishments of the most recent academic week for all the classes that you are taking this semester (or quarter).

See Figure 13–2.

Memorandum

DATE: November 11, 1995

TO: Howard Greenberg, Professor

FROM: Scott Leland   *SL*

RE: Periodic Report—Week of October 30 through November 4, 1995

CE 301—TECHNOLOGICAL ECONOMICS
We discussed Chapter 11 on depreciation and reviewed the solutions to Problem Set No. 6. My reading and problem solutions are complete through Chapter 10. Next week is an exam covering Chapters 7 through 11.

ME 304—STRESS ANALYSIS I
The instructor lectured on fatigue and endurance strength in class and assigned a machine design project to be submitted at the last class meeting. The instructor returned Exam 2 on Tuesday; I scored 83 points. I have not done any work for this class since Exam 2.

ME 305—STRESS ANALYSIS LABORATORY
Our group performed the steel beam fatigue test. Next weekend I will reduce the data and write the report. The test proceeded as we expected, and the data we collected seem reasonable. I received 92 points for the pressure vessel test report submitted on October 26.

ME 302—THERMODYNAMICS
The instructor discussed from a handout the second law of thermodynamics. Although I misplaced my textbook last week, my reading and problem solutions are current. Because we are ahead of schedule, the instructor did not assign any new problems this week.

ECE 232—BASICS OF ELECTRICAL ENGINEERING
We discussed instrumentation for measuring electrical power as covered in Chapter 14. My assignments in this class are up to date. The rough draft of the research assignment that is due next Thursday is completed.

PSY 201—PRINCIPLES OF PSYCHOLOGY
We discussed personal reactions to grief on Monday and Wednesday. The class divided into simulated clinical groups on Friday to discuss our reactions to a stimulus presented by the instructor. A research paper was assigned for submittal by Wednesday, November 16.

FIGURE 13–2
Sample Periodic Report

Critique of the Sample Periodic Report

The periodic report in Figure 13–2 appropriately emphasizes the accomplishments of the student for the period of October 30 through November 4. It includes details for each class, and only critical future events such as exams (i.e., CE 301) and papers (i.e., PSY 201) are discussed. Events before October 30 and after November 4 are not mentioned unless they relate to an event during the reporting period (i.e., ECE 232).

If this memo continued on a second page, the reader would be alerted to a continuation page by the inclusion of *continued* at the bottom of the first page. Also, the top of the second page would alert the reader to the continuation of this first page. If the memo included more than two pages, the pages should be numbered as a fraction to indicate the total number of pages. For example, page 3 of a four-page memo should read 3/4. This fraction ordinarily appears in the footer but sometimes appears in the header.

Sample Assignment—Progress Report

Write a progress report to a client concerning your progress for a project for which you are responsible.

See Figure 13–3 on p. 189

Critique of the Sample Progress Report

The progress report of one project in Figure 13–3 is much broader in scope and more detailed than the periodic report (see Figure 13–2). It discusses specific details of the accomplishments as well as the problems encountered. It includes price (i.e., bathroom fixtures), quality of workmanship (i.e., foundation), and anticipated completion dates (i.e., foundation, bathroom fixtures, and miscellaneous). It also includes start dates and delivery dates (i.e., foundation, framing, and bathroom fixtures), and unavailability of parts specified in the original design and resolution to this problem (i.e., bathroom fixtures).

Sample Assignment—Trip Report

Write a trip report to one of your instructors for a recent field trip that you took. Suggest future student or class activities that could be a result of this trip.

See Figure 13–4 on p. 190.

Critique of the Sample Trip Report

The trip report in Figure 13–4 appropriately begins with a statement of the facility, location and date of the trip, and the function of this facility. However, the writer assumes the reader understands the purpose of this trip, which, inappropriately, is not indicated. For example, this purpose could have been to study the chemical treatment performed at the facility, structural design, or management.

ACON CONSTRUCTION COMPANY
427 NOBLE STREET
MYCITY, HR 12321
(102) 444-5555

March 24, 1995

Dr. Shafiq Ahmad
660 Bluechip Way
Oro City, TH 12321

RE: Progress Report–Construction of Two-Story Residential House
        1280 Gneiss Street, Oro City, TH

Dear Dr. Ahmad:

The status of construction is as follows:

FOUNDATION
Ace Foundations poured the concrete for the foundation on Friday, March 3. As a result of cracking of the foundation caused by the unexpected cold weather, Ace dug up the North side of the foundation and erected new forms on Friday, March 17, and will return to the site on Thursday, March 30, to repour the concrete. This operation requires approximately 2 days. Ace is prepared to work on the weekend, if necessary, for our convenience.

FRAMING
The framing operation is scheduled to begin on Monday, April 3. All primary structural members have been delivered to our site. No problems are anticipated.

BATHROOM FIXTURES
We ordered the bathroom fixtures on February 1. Because of temporary unavailability, the manufacturer substituted the brass faucet that you specified for the master bath with a similar faucet with pearl handles. I have enclosed the brochure for the substituted faucet. If this faucet is not acceptable, please contact me so that I may provide you with another alternative. This change does not affect price or delivery time for the fixtures. The estimated delivery date is June 1.

MISCELLANEOUS
We ordered the electrical and plumbing hardware, which will be delivered in 4 to 6 weeks. Everything else is progressing on schedule.

Please call me if you have any questions.

Very truly yours,

David Garcia

David Garcia, Project Manager
Acon Construction Company

FIGURE 13–3
Progress Report

Memorandum

DATE: May 16, 1995

TO: Julie Wei, Professor

FROM: Kimberly Lane *KL*

SUBJECT: Water Factory 21 Field Trip

I visited the Water Factory 21 municipal water treatment plant in Huntington Beach, California, on Saturday, May 13, 1995. This facility services the Orange County area. In addition to supplying the municipal water to the community, it injects water underground to prevent intrusion of sea water into the ground water supply.

On my arrival at 10:00 a.m., the tour of the treatment facility began. I observed the treatment process in sequence as follows: screening to remove debris, chemical precipitation, flocculation, sedimentation, recarbonation, granular media filtration, disinfection, and reverse osmosis. The reverse osmosis treatment at this facility is known worldwide for its large treatment capacity and advanced technology.

This trip has inspired the idea for a senior project. A student could investigate the feasibility of constructing a desalinization plant near the ocean that uses reverse osmosis treatment. An area that is presently experiencing a drought, such as the City of Santa Barbara, would be excellent for consideration.

FIGURE 13–4
Sample Trip Report

Appropriately, the second paragraph discusses the events of the trip: an observation of the chemical treatment performed at the facility. The purpose of this trip, which should have been included in the opening paragraph, can now be presumed by the reader.

The closing paragraph appropriately suggests a future activity that could be a result of this trip.

KEY CONCEPTS

- Periodic, progress, and trip reports are your opportunity to communicate regularly with supervisors, managers, and clients.
- Periodic reports include sufficient details so that readers can monitor accomplishments during the reporting period and reassign personnel, if necessary.
- Progress reports include sufficient details so that readers can understand specific accomplishments and problems concerning quality of workmanship, time for completion, and cost.
- Trip reports are an opportunity for you to demonstrate your ability to foresee beneficial future actions for your organization.

STUDENT ASSIGNMENT

1. Write a periodic report to your supervisor at work for all tasks for which you are accountable. Include your accomplishments and setbacks during the last calendar month.

2. Write a progress report to your instructor. Include your progress for one class this semester (quarter) that includes a lecture and laboratory. Use headings such as *lecture*, *reading*, *homework problems*, *laboratory*, and *laboratory reports*.

3. Write a trip report to your supervisor for a recent business trip or to the instructor of one of your classes for a recent field trip that you took. Include desired future actions or suggested class assignments or projects that can be a result of this trip. If you have not made any business or field trips recently, you may write a trip report to any interested instructor for an off-campus event you recently attended that has educational or professional value. Suggest a class assignment or project that could be a result from this trip.

CHAPTER 14

Personnel Performance and Salary Reviews

Employees are encouraged to mature professionally with periodic personnel performance reviews.

Employees are the most important resource of an organization. Their performance on the job can make the difference between making a profit or sustaining a loss and can result in completion of a project that satisfies its intended use. It is the responsibility of supervisors to evaluate these employees' performance to help them develop their capabilities to the fullest. This is achieved with personnel performance reviews.

PURPOSE AND AUDIENCE

After you have worked for several years as a professional, other technical personnel will probably be assigned to assist you in performing your responsibilities. You, as their supervisor, will evaluate these employees periodically for the following reasons:

   1. To help these employees improve their work performance: Evaluations are usually reviewed in a confidential meeting between the supervisor and employee. These written evaluations are typically called *performance reviews* (some organizations call them performance appraisals) and become part of the employee's personnel file. The administrations of large, medium, and some small organizations usually provide standard forms to supervisors to facilitate the review procedure (see Figure 14–1 for a sample). These standard forms usually have rating systems for each item included in the evaluation. Less formal organizations rely on memos written by the supervisor to administration.

   2. To help administrators determine salary increases for employees: In many private organizations, the supervisor submits a recommendation to the administration for salary increases with these performance reviews. Sometimes, the administration determines the allocation of the available funds for salary increases based on the performance reviews only. These recommended amounts for salary increases are usually held confidential between the supervisor and administrator, and the employees are not privileged to know this information.

Government agencies and some medium and large companies have predetermined salary schedules whereby each employee's salary is determined from a matrix of employment classifications of increased responsibility (e.g., assistant researcher, researcher, senior researcher) versus steps within each classification. Each employee systematically progresses between steps within a classification unless a step is jumped (company personnel regulations permitting) or the employee is given a promotion to the next higher classification. With predetermined salary schedules, an employee has no guarantee of progress to the next step at each performance review. The salaries in these schedules are periodically adjusted to compensate for inflation.

When limited funds are available for salary increases, employees are sometimes ranked based on the value of their contributions to the organization. Then, to recommend or determine salary increases, these rankings are compared to the present salaries of these employees. Also, ranking of employees is common when professional staffs need to be reduced because of diminished work loads or sales.

3. To justify promotions of meritorious employees: In addition to periodic increases in salary, competent performance can be rewarded by an increase in responsibility. This increase in responsibility generally requires a promotion to the next higher classification. A review of an employee's personnel file is usually the basis for this promotion. The employee is usually notified of the promotion during a performance review and receives a notification or memo from administration to acknowledge this change of classification.

Each organization has its own calendar for performance reviews and recommendations for salary increases:

- All employees can receive performance reviews and recommendations for salary increases at the same time: commonly January 1, July 1, or both.
- Each employee can receive performance reviews and recommendations for a salary increase on the anniversary of the start of employment with the organization.

However, modern management techniques encourage separating performance reviews from recommendations for salary increases to emphasize the importance of the performance reviews. To accomplish this, many organizations grant salary increases to all employees at the same time (this simplifies budgetary considerations) and give performance reviews on the anniversary of each employee's start of employment.

PERSONNEL PERFORMANCE REVIEW - TECHNICAL STAFF

Name:_____ Team:_____ Date Prepared:_____

Job Description:_____

Supervisor:_____ Title:_____

SECTION ONE

CURRENT ASSIGNMENT (Describe work being performed)

OTHER ASSIGNMENTS (Describe other major assignments during last six months not presently being performed)

NOTEWORTHY ACHIEVEMENTS DURING LAST SIX MONTHS

SECTION TWO - Factor Measurements

FACTOR MEASUREMENTS (Please circle)

10 - Superior
7 - More Than Satisfactory
5 - Satisfactory
3 - Improvement Needed
1 - Unsatisfactory

1. *JOB COMPETENCE*   10 9 8 7 6 5 4 3 2 1

Basic job knowledge and skills, understanding of specific job duties and familiarity with other job functions. Growth in capabilities and competency areas. Originality of approach to problems. Amount of self-directed effort.

FIGURE 14-1
Sample Personnel Performance Review (pp.195-197)

2. *WORK QUALITY*   10   9   8   7   6   5   4   3   2   1

Quality of results - accuracy, neatness, thoroughness.

---

3. *INITIATIVE*   10   9   8   7   6   5   4   3   2   1

---

4. *DEPENDABILITY*   10   9   8   7   6   5   4   3   2   1

Extent to which employee accepts and follows through on assignments, record of seeing jobs through to completion.

---

5. *WORK HABITS*   10   9   8   7   6   5   4   3   2   1

Attendance and punctuality, observation of policy and procedures, care of equipment, etc.

---

6. *RELATIONS WITH OTHERS*   10   9   8   7   6   5   4   3   2   1

Extent to which employee works effectively with fellow employees and with clients.

---

7. *ATTITUDE*   10   9   8   7   6   5   4   3   2   1

Flexibility in types of projects willing to undertake, willingness to perform under pressure, interest in new fields, cooperation, dedication and willingness to spend extra time when required, willingness to perform in tedious as well as interesting work.

---

8. *PROJECT MANAGEMENT*   10   9   8   7   6   5   4   3   2   1

Leadership qualities, effectiveness in decision making, technical and financial project performance.

---

9. *MARKETING PERFORMANCE* 10   9   8   7   6   5   4   3   2   1

Client relations, initiative, success in attracting new clients, ability to follow through on marketing leads, effective participation, ability to arrive at realistic budget and scheduling.

---

10. *OTHER*   10   9   8   7   6   5   4   3   2   1

(Specify)

FIGURE 14-1 (*continued*)

SECTION THREE - Supervisor's Comments

(Include description of employee's major strengths, areas of weakness, potential for future growth, etc.)

SUPERVISOR'S OVERALL EVALUATION (Check One)

○   Superior   ○   Improvement Needed
○   More Than Satisfactory   ○   Unsatisfactory
○   Satisfactory

SECTION FOUR - Improvement

ACTION TO BE TAKEN BY SUPERVISOR TO ASSIST EMPLOYEE IN IMPROVING PERFORMANCE (Indicate time frame)

ACTION TO BE TAKEN BY EMPLOYEE TOWARDS IMPROVING OWN PERFORMANCE (Indicate time frame)

SECTION FIVE - Employee Comments

SECTION SIX - Signatures

Supervisor's Signature:_____ Title: _____

*Employee's Signature:_____ Date:_____

Admin. Vice President's Signature:_____

*Your signature indicates neither agreement nor disagreement with this evaluation, but it does indicate the evaluation has been reviewed by you.

When employees receive notifications for salary increases without performance reviews, these employees should be given informal performance reviews to foster improved working relationships between them and their management.

CONTENT

Performance reviews are critical analyses of employees' performances during the review period. Progress and growth since the previous review are evaluated. Strengths and weaknesses and possibilities for growth within the company are also reviewed. Judgments concerning work performance (e.g., work efficiency, technical accuracy, and interpersonal and communication skills) are justified with examples.

Many organizations encourage active employee participation in the performance review process by including a self-assessment by the employee. This is completed by the employee in response to either

- Goals and objectives that were established at the most recent review
- The supervisor's comments and evaluations for this review period

Goals and objectives may be established for the next review period.

Performance Reviews

Performance reviews are the primary indicators to employees of their potential for long-term growth with the organization, and therefore supervisors should take them seriously. When standard forms are not provided, the following items should be addressed:

1. *Achievements since last review:* Professional accomplishments outside the place of employment, such as completion of classes and certification in a new discipline.

2. *Analysis of recent projects:* A summary of the success of the employee's performance of projects worked on from the viewpoint of completion on schedule and within budget and quality (e.g., accuracy, thoroughness). Problems that were encountered in this performance and their resolution by the employee may be discussed.

3. *Strengths (weaknesses):* Competence, judgment, quality of work performed, dependability, interpersonal relations with co-workers and clients, initiative, work ethic (e.g., attitude toward work, attendance and punctuality, care of equipment), and management skills (projects, time, and people).

4. *Possibilities for growth:* Possibilities for increased responsibilities and advancement in the organization during the next review period. This is based on the skills of the employee matched with the outlook for changes within the organization.

5. *Recommendations:* Specific steps to increase the value of the employee to the organization (e.g., learning a new skill, improving interpersonal relations with shop personnel).

Sometimes the performance of an employee does not meet the minimum standard for one or more of the performance criteria of the organization. The reviewer must be tactful in preparing and discussing this employee's review, because this situation may cause irreparable harm to the organization by causing resentment by the employee. When this situation arises, it is important to emphasize the employee's strengths while recommending courses of action to improve performance of the employee. The personnel department of most organizations can assist you in preparing this performance review in a manner that will prevent alienating the employee from the organization and, at the same time, protect the organization from a future lawsuit if the employee's performance does not improve and the organization takes action against the employee.

Sometimes the performance of an employee merits dismissal from the organization. Employees rightfully have significant protection under the law, and the dismissing organization may be required to defend its action in court if the employee is dismissed improperly. When the performance of an employee requires dismissal, this action should be performed by the personnel department with the facts and data that have been documented by the supervisor.

Recommendations for Raises

When recommendations to the personnel department for employee raises occur at the same time for all employees, and standard forms are not used, these recommendations should include a justification to administration for the raise that is recommended for each employee. These justifications should discuss each employee's strengths and weaknesses. When a limited amount of money is allocated for raises, the total of the recommended raises should be shown. Names of employees can be used as headings.

When each employee is recommended for a raise at a time different from other employees, and standard forms are not used, it is expedient to write memos to administration that summarize and make current these employees' most recent performance reviews.

Sample of Strengths and Weaknesses Section from the Performance Review of a Technical Marketing Representative

Stanley Cooper has an innovative approach for marketing our new line of XL-5 relay systems. The effectiveness of this technique has increased the sales of our department by 18% compared to sales for the same quarter of last year.

Stan is adept at building good working relationships with new clients. However, in several instances he was not responsive to the needs of our existing clients. In fact, after he ignored requests from Michaels Manufacturing to investigate the possibility of reducing the width of our G-5 relays, they did not renew their contract with us.

The technical personnel enjoy working with Stan and appreciate his ability to foresee potential applications for new products. Our sales to Barr Aircraft of almost $300,000 per year for G-7 relays is a result of Stan's foresight.

Although Stan is occasionally overly aggressive with his co-workers when promoting his ideas, he communicates well with others in the department, and the junior marketing representatives find that he is a valuable resource person.

Critique of Sample Strengths Section From a Performance Review

This sample is packed with information, data, and examples. Because of the clarity of thought and detail included, this section of the review appropriately sets up the reviewer to discuss the possibility for growth of the employee and, in the subsequent portions of this memo, recommendations for improvement.

Although this review includes two negative comments, it is generally very positive. Unfortunately, employees tend to place greater emphasis on the negative comments that are included and overlook the positive comments. To minimize the psychological impact of the negative comments, this reviewer

- Indicates an omission (i.e., "after he ignored requests") rather than a failure on the part of the employee
- Minimizes the impact of a negative comment ("gets overly aggressive") by introducing it with the transition word *although*, followed with a positive strength

Because these comments are specific and expressed diplomatically, the employee should accept them as constructive criticism.

SAMPLE RECOMMENDATIONS FOR RAISES

Figures 14–2 and 14–3 are samples of recommendations for raises. A discussion of the sample follows.

Critique of Sample Recommendation 1

The introductory paragraph in Figure 14–2 appropriately states the purpose of, and personnel evaluated in, the memo, and the total funds available for salary increases. This prepares the reader for reading the text below.

Headings, not included in this sample, would make the memo easier to read and remember.

The strengths that justify a salary increase for each employee are appropriately discussed before this increase is recommended. However, specific actions that validate the judgment of these strengths, such as successful completion of a project with a limited budget, should be included.

Bridge Design, Inc.

MEMORANDUM

DATE: March 1, 1995

TO: Michael Roden
          Personnel

FROM: P. M. Aster *PMA*

Subject: Salary increases

Enclosed are my recommendations for the total distribution of $600 per month in salary increases for Group 235 which includes Daniel Benjamin, Julia Loren, and Douglas Marquez.

Daniel Benjamin works well with others in a group setting. His communication skills are good and he demonstrates adequate knowledge of testing procedures. I recommend that he receive a raise of $225 per month.

Julia Loren did a good job preparing the analysis. She works well in a group environment and demonstrates that she is responsible. I recommend that she receive a raise of $200 per month.

Douglas Marquez was enthusiastic and helped the other group members prepare the various segments of the project. He works well as a team player and is willing to take responsibility. I recommend that he receive a raise of $175 per month.

Please notify me by March 8 that you have received my memo. If you have any questions or comments, please call me.

FIGURE 14–2
Sample Recommendation for Raises 1

Because strengths are likely to continue, this sample appropriately discusses them in the present tense. Completed actions, however, are appropriately addressed in the past tense.

The closing paragraph requests action by a specific date because there is a deadline.

The statement "If you have any questions or comments, please call me" in the closing paragraph is acceptable in formal business correspondence; however, it is trite in a memo from one employee to another in an organization. It therefore should be deleted.

Critique of Sample Recommendation 2

The sample in Figure 14–3 appropriately addresses the "Re" as "Personnel Evaluation" rather than "Salary Increases," the subject of the memo. Inappropriately, however, an introductory paragraph has not been included. Therefore, the reader is not guided into the text of the memo.

The headings in this sample alert the reader for the text that follows.

As in the previous sample, the strengths that justify a salary increase for each employee are appropriately discussed. Also, specific actions are discussed.

Each employee in this sample is curtly recommended for a salary increase before the strengths of that employee are discussed. More appropriately, the justification for each increase should be discussed before the recommendation is made.

KEY CONCEPTS

- Personnel are the most important resource an organization has. Personnel reviews are the most efficient method of developing this resource.

- When salary increases are separated from performance reviews, the supervisor should take the opportunity to have an informal performance review when the employee is given a salary increase to foster a more personal relationship.

- Sometimes an employee must be given a mediocre or negative performance review, or a salary increase that is less than expected. A tactful personal discussion with the employee about the employee's performance or justification for the less than expected raise can help eliminate resentment by the employee.

STUDENT ASSIGNMENT

1. Select a class that required group participation of at least three to four students to complete a common project or laboratory experiment. Assume that the other students in the group report to you and that you are responsible

PS STRUCTURES

MEMO

Date: November 11, 1995

To: Peter Stuyves

From: Lenore Fliedner, P.E. *LF*

Re: Personnel Evaluation

JULIE POLACK
I recommend a raise of $225 per month for Julie Polack. Julie demonstrates
initiative and a willingness to work on any project to which she is assigned.
She is thorough in her design and careful in her methods. She works well
with associates and listens to their suggestions. She demonstrates a good
grasp of the essential design principles and leadership capability.

LARRY ZELIG
I recommend a raise of $195 per month for Larry Zelig. Larry did an out-
standing job on the last project in preparing the test specification for review
and approval by the owner. He complied with changes that were requested
and promptly returned them for final approval. Although his interaction with
associates is minimal, he works well with others and is willing to help when
necessary.

ILENE LEONARD
I recommend a raise of $180 per month for Ilene Leonard. Ilene has been
helpful in document preparation and review. On the last project, she aided
Larry Zelig in review of the specification in preparation for drawing submittal. She also prepared and submitted the bid documents.

FIGURE 14–3
Sample Recommendation for Raises 2

for their performance reviews and recommendations for raises. Select names of famous personalities from movies, TV, rock, comic strips, etc. for your subjects, and do one of the following in the form of a memo:

   a. Evaluate the contribution of each member of your group, and write a recommendation for raises to administration. Assume that you have a maximum of $750 per month that can be distributed for raises.

   b. Write a performance review for one of the members of your group.

2. Select a job that you presently work at or worked at recently. Assume that three or four other employees that you work (or worked) with report to you and that you are responsible for their performance reviews and recommendations for raises. Select names of famous personalities from movies, TV, rock, comic strips, etc. for your subjects, and do one of the following in the form of a memo:

   a. Evaluate the contribution of each employee, and write a recommendation for raises to administration. Assume that you have a maximum of $800 per month that can be distributed for raises.

   b. Write a performance review for a technical employee (e.g., drafter, technician).

   c. Write a performance review for an administrative employee (e.g., expediter, administrative aide).

   d. Write a performance review for a secretary.

   e. Write a performance review for an employee who is not fulfilling the duties of his or her job description.

CHAPTER 15

Bids and Proposals

Contracts are awarded to competing organizations by submittal of bids and proposals.

Reports typically describe what is either planned or has already been accomplished. On the other hand, bids and proposals describe what is proposed to a potential client before being awarded a contract.

The award of contracts is based on the submittal of bids and proposals. These bids and proposals usually compete with other organizations that have capabilities similar to yours. The award of these contracts determines the financial success of most professional organizations in the private sector and some universities and research organizations.

The objective of bids and proposals is to convince potential clients of your ability to perform the job at least as competently as your competitors, at the stated price. Therefore, bids and proposals need to be persuasive in their presentations by being thorough and accurate and by presenting a clear picture of the outcome of the work that is proposed by including its details. Also, it is important that the presentation of your bids and proposals is attractive and captures the reader's attention.

Invitations to submit bids and proposals can be found in publications such as the *Engineering News Record*, the *Federal Register*, and the classified section of major newspapers. Information concerning the documentation required for a response to these invitations can be obtained by replying to these advertisements.

PURPOSE AND AUDIENCE

When a company or a government agency lacks the internal capability to accomplish a specific task, it usually seeks an outside organization to accomplish this task. A statement of this task, in the format of a specification (a detailed account of an organization's requirements; see Chapter 16 for more information), is usually sent to a qualified organization (as determined by a prior technical audit or other selective process) with "Instructions to Bidders," which explains the procedure for responding.

Bids Compared With Proposals

A bid responds to a Request for Quotation (RFQ). An RFQ is released when a company or government agency has a specific need with a well-defined solution and requests a price for complying with that need from bidders. For example, an organization would release an RFQ for the design and manufacture of a two-passenger electric vehicle to be used only for on-site inspection of construction sites.

Most often, bidders are requested to submit "Comments and Exceptions" or "Deviations" (sections of the specification that the bidder cannot comply with) to the specification with the bid price. When the comments and exceptions or deviations are acceptable and do not disqualify the bidder from the bidding, the competitive bid price is the basis for the award of the contract.

Usually, a proposal responds to a Request for Proposal (RFP). An RFP is released when the company or government agency has a specific need with a solution that is not well defined and requests both a design and price for complying with that need from proposers. For example, an RFP would be released for the design and construction of a transportation system for commuters from the central business district to the outlying bedroom communities. The proposal that meets this need may be a light-rail transit or a high-speed freeway. In the proposal, the proposer needs to convince the company or government agency of the desirability and technical feasibility of the proposed solution and to include the proposal price. The award of the contract is based on the suitability of the solution for its intended purpose, cost-effectiveness of the proposed solution, the funds available for the project, and the technical competence of the bidder to perform as promised.

An unsolicited proposal is submitted when there is a prior relationship with the department, company, or government agency to whom the proposal is submitted and the proposer perceives a need. An unsolicited proposal may be used in one of the following situations:

- To change the scope of an existing contract with a client.

- To perform a service to another department within your own company or government agency.

- To increase the capability of your own department by procuring special tools or equipment, or by hiring specially trained personnel. These proposals are usually submitted to administration for approval.

A proposal is sometimes a response to a verbal request from a company or government agency. This proposal is very similar to a response to an RFP except that the proposer has a greater latitude in the development of the solution because there is no formal specification that needs to be complied with, and there is no specified format for the response because no instructions to bidders are given.

The proposer needs to use considerable care in accurately identifying the need of the company or government agency before proposing a solution.

A grant proposal is a request for research funds to a government agency, philanthropic foundation, or other organization. A grant proposal is commonly made by universities and research organizations. The award of the funds for these grants is competitive, but the research topics that can be proposed for these grants may be broad in scope, and therefore, their proposed solutions are noncompetitive. The proposer of a grant proposal must convince the granting organization of the worthiness of the research that is proposed.

Mailings to concerned organizations and notices in professional publications publicize the availability of these funds for grant proposals. The granting organizations generally have specific formats for submitting these proposals and deadlines that need to be adhered to. These formats are usually very similar in content to the headings listed in "Proposals," later in this chapter.

Grant proposals for research should state the gap in knowledge the research is intended to fill, state a hypothesis, and give the reasons for holding that hypothesis.

CONTENTS

A bid or proposal becomes part of the legal contract on acceptance by the company or government agency to whom it is submitted. Therefore, a bidder or proposer must evaluate the requirements thoroughly, because noncompliance with the bid or proposal after the award of the contract is a breach of contract and may become the subject to a lawsuit. It is important to comply with the instructions to bidders in responding to RFQs and RFPs to avoid disqualification in the bidding process. The submittal of the bid or proposal must meet the deadline to the minute to be considered for the award.

Typically, after technical personnel review the bid or proposal submittals, they make recommendations for the award of the contract to managerial and financial personnel who make the final decision. Therefore, it is important for the bid or proposal package to convince persons of different backgrounds of the capability of the bidder or proposer to satisfy the needs of the awarding organization.

Frequently, the expectations of the releasing organization are not clearly specified in the RFQ or RFP. When this occurs, these organizations are usually receptive to questions you may have concerning unclear or ambiguous requirements, and they may revise their document to prevent misunderstandings.

Bids

The documentation required in response to RFQs (bids) usually includes the bid price, comments and exceptions or deviations, and other materials that may be requested—sometimes referred to as *boilerplate* by the bidding organization because of its bulk (e.g., qualifications of personnel who will be assigned to the project or a list of similar projects completed).

For each section of the specification used for bidding, the instructions to bidders usually allow only one of the following four responses with an explanation in the comments and exceptions or deviations:

- "Will Comply." The bidder will comply with the section. For example, the specification may read "The room shall be 15 ft wide by 18 ft long," and the bidder intends to build the room to that size.

- "No Comment." This section is informational only, and no additional response is required. For example, the specification may read, "This vehicle will be used for competitive racing," and no specific requirements need to be complied with. However, later sections in the specification probably will address the performance characteristics of the vehicle that cause it to perform as a racing vehicle.

- "Comment." The bidder intends to comply with the section, but there is a limitation or recommendation. For example, the specification may read, "The rotor shall have an 18.25-in. diameter."

  The bidder might add the following limitation, "Presently, as the result of a labor strike, steel of the appropriate diameter for the rotors is not available in this country. Unless this strike is settled before fabrication, delivery may be delayed up to 2 months."

  Or the bidder might add the following recommendation, "Buildup of allowable tolerances and eccentricity that are due to wear may eventually cause interference between the rotor and its housing. Therefore, an 18.125-in.-diameter rotor is recommended."

- "Exception." The bidder will not comply with the section, either because it is not within the technical capability of the bidder, or it is not economically competitive to do so.

  For example, the specification may read "The 90° bends in the steel shall not exceed a 1/4-in. radius." The bidder may not want to comply with the requirements for this small radius because of a limitation of the capability of its forming equipment or the additional expense, and therefore the bidder responds, "Exception: 90° bends in the steel shall not exceed a 1/2-in. radius."

  Exceptions place the bidder in a competitively disadvantageous situation and should be used only when necessary.

The bid price is usually included on a form provided by the awarding organization. Otherwise, it can be included in a cover letter.

Proposals

The documentation for proposals must be convincing to the reader. A letter of transmittal, a professional-looking cover and title page, a permanent binding, and clear graphics should be included. Proposals frequently are printed by the graphics or publications department of the proposing organization.

Proposals usually include the following sections (see Figure 15–1 on p. 210):

1. *Summary:* a brief description of the problem and the proposed solution, the cost, and a list of the qualifications of the persons to be assigned to perform the work.

2. *Statement of the problem:* a detailed account of the problem.

3. *Background:* conditions that led to the problem and previous attempts to solve it.

4. *Scope:* a statement of what products and services will be provided by the proposer to solve the problem, and what is intended to be accomplished. Any work that is specifically excluded may be mentioned.

5. *Proposed solution:* a detailed account. This section is the main part of the report. Include sufficient graphics, data, and calculations for the reader to understand what is proposed and its technical feasibility.

6. *Advantages and disadvantages of the proposed solution:* a statement of the problems that proposed solution will resolve. Any problems that will not be affected or conditions that may be adversely affected should be addressed to avoid misunderstandings and potential lawsuits.

7. *Alternative solutions:* other possible solutions that were considered for solving the problem and the reasons for eliminating them.

8. *Schedule:* a schedule or Gantt chart (see Figure 13–1) showing critical completion dates.

9. *Price:* an accounting of the proposed prices for services, materials, equipment, and expenses.

10. *Documentation:* drawings, reports, and other documents that will be submitted to the awarding organization, and the schedule for their submittal.

11. *Appendix:* qualification of personnel and facilities. This is a statement of the capability to perform the proposed work. This usually includes résumés of key personnel, similar projects completed, and an inventory of equipment. An annual report may be included.

Sample Scope Section from a Proposal for a Storage Shed

The purpose of this proposal is to provide Joel Elene, Inc., with a storage shed for tools and heavy equipment. A sketch of the proposed storage shed is shown in Figure 1, but the final design will not be completed until Joel Elene, Inc., gives approval.

Critique Sample Scope Section From a Proposal

The opening phrase, "The purpose of this proposal is to provide," is wordy and implies that the *proposal* (rather than the proposer) will perform the activities indicated in this proposal. Also, the word *provide* is vague. Does it mean design, build, loan, or deliver? This needs to be clarified.

Rather, this section could begin with "A proposed storage shed for the tools and heavy equipment will be designed and built for Joel Elene, Inc., as shown in Figure 1." This statement makes clear that the product to be designed and built is demonstrated in Figure 1 (not shown in this text). This eliminates any misunderstandings or misconceptions that may occur.

FIGURE 15–1
Sections of a Proposal

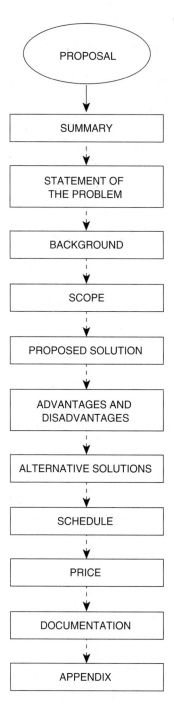

The final phrase concerning approval, "but the final design will not be completed until Joel Elene, Inc., gives approval," appropriately is a condition in a specification for completion of the design but does not belong in a scope because it does not address the items intended to be accomplished.

SAMPLE RESPONSES TO A REQUEST FOR PROPOSAL

Figures 15–2 and 15–3 illustrate responses from two different designer/builders to an RFP from a university to design and constuct a student parking lot or structure with 400 spaces.

Critique of RFP Response—Sample 1

The sample in Figure 15–2 is organized well and effectively uses figures to supplement the text and tables to show the proposed schedule and price. Although the sample has some problems, as discussed later, the text appropriately builds on information already presented.

The tone of the summary sounds more like an introduction. The opening sentence could be "A 400-space parking lot is proposed on the northwest corner of Temple Avenue for California Polytechnic University, Pomona. The work will be completed by September 25, 1996.

The summary says that the work will be supervised by J. A. Halverson and Associates. It is not clear if the proposal is merely for supervision or includes design and construction, as stated in the sample assignment.

A statement of the problem is appropriately included at the beginning of the report. The opening sentence concerning student surveys would be more appropriately included in the background.

The background appropriately includes the history of the problem in a qualitative as well as quantitative manner.

This sample does not have a scope, which is essential so the reader knows what the proposer intends to be responsible for.

The proposed solution includes the details and a figure of the proposed parking facility in text format supplemented by a plan of the design details. This makes it easy for the reader to visualize and understand.

The advantages and disadvantages of the proposed solution are included. The advantages of the proposed solution and the price are the primary criteria for awarding the contract. Yet, this discussion lacks sufficient detail for the reader to validate the proposer's claims concerning these advantages and disadvantages.

This sample appropriately includes the advantages of alternative solutions and also discusses the reasons they were not selected.

Qualification of personnel and facilities is appropriately included to demonstrate the competence of the proposer to perform as promised. Appropriately, this is included in the body of the proposal rather than, as sometimes is done, in the appendix. The figures referenced in Sample 1 (Figure 15-2) are not shown. In an actual report, these figures *would* be presented.

PROPOSAL FOR CONSTRUCTION OF A 400-SPACE PARKING LOT
AT CALIFORNIA POLYTECHNIC UNIVERSITY

by J. A. Halverson and Associates
1243 1st Avenue, Pomona, CA 92345

October 12, 1995

FIGURE 15–2
Response to an RFP—Sample 1 (pp. 212–217)

Table of Contents

Summary

This proposal is for California Polytechnic University, Pomona to pay $244,200 for the construction of a 400-space parking lot on the northwest corner of Temple Avenue. The work will be supervised by J. A. Halverson and Associates during the 1995 fiscal year and is to be completed by September 25, 1995.

Statement of the Problem

In a 1995 survey of students' concerns at California Polytechnic University, lack of parking space was the most frequently mentioned problem. The ratio of spaces to parking permits is 1:4.5. This ratio is significantly lower than the 1:4 ratio recommended by the California State University Department of Transportation. The shortage of parking space causes students to be late for their classes, and creates traffic hazards in the parking lots and on feeder streets when students compete for spaces. This shortage occurs on school days from 9 a.m. to 1 p.m. and during the first week of each quarter, when students are registering for classes.

Background

Cal Poly was founded in 1940 as a men's agricultural school on land donated by J. R. Kellogg. The site had been an Arabian horse ranch in a rural area of Los Angeles County. Initially, all students lived on campus. As the surrounding population grew, so did the school population: from 186 students in 1940 to 14, 637 in 1995. Currently, the student housing on campus can accommodate 856 students; the other students commute an average of 8.5 miles.

In 1960, a 1500-space parking lot was constructed between the campus and Kellogg Drive (see (Figure 1). In 1976, an 800-space parking lot was built between the athletic fields and Temple Avenue. Studies by the traffic engineering department on campus demonstrate that the parking lots reach 100 percent capacity at the peak hours which occur on school days between 9 a.m. and 1 p.m., and during the weeks of registration between 7:30 a.m. and 4:30 p.m. According to campus police records, parking citations during these periods increase by 62 percent. Also, the Pomona Police Department reports higher rates of parking and moving violations, and traffic accidents on streets surrounding the campus during these times.

1

FIGURE 15–2
(*continued*)

Proposed Solution

The construction of a 400-space parking lot (see Figure 2) will reduce the ratio of parking spaces to permits to 1:3.8, which is below the recommended ratio. This ratio will remain within the recommended limits for the next 3 years if projected enrollments are not exceeded.

The lot will be located on the west side of Temple Avenue at the intersection of University Avenue. This site is currently being used as a sheep pasture by the school's Agriculture department. The entrance to the lot will be at the signal-controlled intersection north of the flood control channel. The lot will require minimal grading. Approximately 46,000 cu yds. of fill will be used and the lot will be constructed of an aggregate, porous base, and an asphaltic finished surface. The driveway entrance will be concrete, a 6-ft chain link fence will enclose the area, and a steel bar gate will protect the entrance. An existing water system will be converted to a landscape irrigation system along the west side of Temple Avenue.

Advantages and Disadvantages of the Proposed Solution

The primary advantage of building a parking lot on this site is the immediate reduction of the space-to-permit ratio. Other advantages are low cost for construction, maintenance, and security, and preserving opportunities for future building needs. The primary disadvantage of a parking lot in this location is the loss of a scenic sheep pasture.

Alternative Solutions

Alternative solutions might be to use the existing parking spaces and have an intensified program to encourage car pooling, and using mass transportation. This approach has been used on other Cal State campuses; however, the results have been discouraging. Currently only 5 percent of the student body and 7 percent of the school staff throughout the Cal State System use public transportation to commute to school. They complain of infrequent service and lack of safety. Also, after an intensive car pooling campaign on campus, a 1994 study by the Cal Poly Traffic Engineering department found the average number of riders per car to be 1.03.

2

Multistory parking structures is another alternative that has been used on other campuses. This solution has the advantage of increasing the capacity of existing lots near campus and preserving open space. However, the structures are more expensive and require more construction time to build than surface lots. In addition to more expensive construction costs, parking structures require additional security precautions as a result of reduced visibility, and they alter the visual impact of the campus.

Qualification of Personnel and Facilities

J. A. Halverson and Associates has been in business for 25 years and employs four California-registered professional engineers for project design and supervision. Our earth-moving and pavement crews are members of Construction Union Local 45. Our main office is at 1243 1st Avenue, Pomona. We currently have paving contracts with the California Department of Transportation and the City of Pomona.

Schedule

The time necessary for completion of the project are:

Planning/Approval	4 weeks
Survey	1
Earthwork	1
Concrete/Asphalt work	1
Fence and Landscape	1
Total	**8 weeks**

3

FIGURE 15–2
(*continued*)

Price

    The proposed costs are as follows:

Survey party	2 hrs	@ $150/hr	$ 300
Earth moving equipment	10 hrs	@ $200/hr	2,000
Concrete/asphalt - 144,400 sq.ft.		@ $1.50/sq ft	216,600
Chain link fence - 16,000 ft		@ $1.25/linear ft	20,000
Steel bar gate - 20 ft			100
Irrigation/landscape			3,200
Professional fee			2,000
TOTAL			**$244,200**

4

Critique of RFP Response—Sample 2

The sample shown in Figure 15–3, similar to Sample 1, is organized well, effectively uses figures to supplement the text and tables to show the proposed schedule and price, and addresses the items that need to be addressed. Lists used throughout this proposal help the reader comprehend the information.

A summary or an abstract, not included in this sample, is essential in a proposal to introduce the reader to the proposed solution. This is a major oversight by the writer.

A statement of the problem, as in Sample 1, is appropriately included at the beginning of the report.

The background in this sample includes considerable data in lieu of the historical account. All these data, when presented in text rather than tables, facilitate understanding, which may be perceived by the reader as being overly quantitative. Therefore, the reader may feel that it is not necessary to read them carefully. When all these data are included, they should be shown in a table that supplements the text for ease of understanding.

The scope appropriately includes the limitations of the proposal.

The proposed solution, as in Sample 1, includes the details of the proposed parking facility in a manner that is easy to visualize and understand, but it should also include a figure of the proposed solution.

Sample 2 appropriately includes significant detail concerning the advantages and disadvantages of the proposal in an easy-to-understand list. The specificity of the statements concerning the disadvantages help prevent any misunderstandings.

A discussion of alternative solutions is included in a list, which, as in Sample 1, appropriately indicates the reasons they were not selected.

This sample, like Sample 1, appropriately includes qualification of personnel and facilities, but it also advantageously includes information concerning projects worked on and awards.

Categories within the lists of project schedule and price breakdown facilitate comprehension by the reader.

Finally, a title page should precede the appendix of this proposal.

The figures referenced in Sample 2 (Figure 15-3) are not shown. In an actual report, these figures *would* be presented.

PROPOSAL
THE DESIGN AND CONSTRUCTION OF A
400-SPACE PARKING FACILITY
AT
CALIFORNIA STATE POLYTECHNIC UNIVERSITY, POMONA

TO:
THE CALIFORNIA UNIVERSITY
BOARD OF DIRECTORS
SACRAMENTO, CALIFORNIA

BY:
G & F CORPORATION
LOS ANGELES, CALIFORNIA

October 12, 1995

FIGURE 15–3
Response to an RFP—Sample 2 (pp. 219–225)

Table of Contents

FIGURE 15–3
(*continued*)

Statement of the Problem

Present parking facilities at California State Polytechnic University, Pomona accommodate 7900 vehicles. Because of the projected growth of the University population, these facilities will be inadequate by 700 parking spaces in the next 5 years.

Currently, a parking facility to accommodate 600 vehicles is under construction and will increase parking capacity to 8500 vehicles. When this new facility is completed, the projected need for parking spaces will still not be met.

On January 3, 1995, university administration requested proposals for the design and construction of a new parking facility with a 400-vehicle capacity. Construction of the facility is intended to be completed by July 1, 1996.

Background

In 1990, the university population needed 5600 parking spaces. However, the parking facilities had a capacity for 6000 vehicles. In 1991 and 1992, the parking space requirements were 5900 and 6200 respectively. In 1992, new parking facilities were constructed which increased parking capacity by 900 vehicles, to 6900. In 1993, parking space requirements increased to 6600, and parking capacity increased by 1000 vehicles, to 7900. In 1994 and 1995, parking space requirements were 7200 and 7800 vehicles respectively, and parking capacity was increased to 8500. Current parking space requirements are 7800 vehicles. Current projects will increase parking capacity to 8500 vehicles by the end of 1996. Because a capacity of 8500 vehicles will not meet projected requirements of 8600 spaces for 1997, new facilities must be constructed by July of 1996.

Scope

The scope of this project includes the planning, design, and construction of a 400-space parking facility. The following limitations apply:

1

1. Parking space use requirements are as follows:
   8 handicap spaces
   2 service vehicle spaces
   390 standard spaces
   0 compact spaces
2. Either an open parking lot or a parking structure is acceptable. If a structure is built, it will include a maximum of two above-ground levels.

Proposed Solution

The proposed parking facility is a 400-space open parking lot located adjacent to Building 143. The paved and landscaped area of the parking lot will require 3.5 acres. Angled parking will be used to restrict traffic flow and increase pedestrian safety. Islands will be landscaped and have a total landscaped area of 0.3 acres. Parking spaces will include 8 handicap spaces, 2 service vehicle spaces, and 390 standard spaces. Vehicles entrances will be located at the four corners of the lot. Entrances on the north side of the lot will be accessible from Rosaria Lane, and entrances on the south side of the lot will be accessible from Joseph Lane. Pedestrian pathways will be built throughout the parking lot; pedestrian entrances will be limited to those shown on the Site Plan. Pedestrians will be restricted to using the pedestrian entrances by landscaping surrounding the perimeter. Lighting, storm drain, signing and striping plans are shown on the Utility and Traffic Control Plan. See the Appendix for the Site Plan and the Utility and Traffic Control Plan.

Advantages of Proposed Solution

Advantages of the proposed solution follow:
1. The proposed location improves the availability of parking facilities within campus. With the construction of this facility, lots will be located on all sides of the academic center of campus. Also, buildings that are not readily accessible from the majority of the existing parking facilities are near the proposed lot.

2

FIGURE 15–3
(*continued*)

2. A parking lot is more aesthetically pleasing than a parking structure and has lower construction, maintenance, and lighting costs.
3. Islands with landscaping enhance aesthetics and reduce the environmental impact of the facility.
4. The vehicle entrance locations increase traffic on streets that do not feed other major parking facilities. Any increase of traffic on the streets that have access to other parking facilities would require additional lanes to increase the capacity.
5. Land to the west of the proposed site is available for future expansion of approximately 700 additional spaces.

Disadvantages of Proposed Solution

Disadvantages of the proposed solution follow:
1. Construction on the proposed site will require an expensive drainage system because hills surround 65 percent of the site.
2. Dust and noise from construction may disturb agricultural studies that are south of the proposed site.

Alternative Solutions

Alternative solutions follow:
1. Construction of a parking lot north of the Administration Building would reduce the cost of grading because the land is relatively flat in the area. However, this construction requires the construction of two 300-foot access roads. The potential cost of construction of these access roads is greater than the potential savings of grading costs. The location of this facility would not service the major academic portion of the campus as effectively as the proposed location.
2. Construction of a parking structure above the existing lot east of the library would require less land. This leaves open land for expansion of future facilities. This site is an excellent location for a parking structure because it services most of the major academic building. This alternative solution is less aesthetically pleasing than the proposed solution to the environment and increases construction, mainte-

nance, and lighting costs. The solution also requires improvements to the existing access streets as a result of increased traffic.

3. Construction of a parking structure at the proposed location would require less open space and leave the unused acreage for expansion of future facilities. This solution, though, would be less aesthetically pleasing to the environment and requires increased construction, maintenance, and lighting costs.

Qualification of Personnel and Facilities

G & F Corporation has been serving the civil engineering and construction needs of its clients since 1953. From a ten-person local operation 42 years ago, G & F has grown into a 35-office, 3700-employee, nationwide corporation.

G & F has designed and constructed parking facilities for companies and agencies such as Bechtel in Norwalk, the City of Los Angeles, Warner Brothers in Burbank, Disneyland, Disneyworld, and the Marriot Corporation, as well as numerous hotels, restaurants, and recreational and industrial facilities.

G & F has received numerous awards such as the Portland Cement Institute's Best Parking Facility Award in 1992, A.S.C.E.'s Economics and Quality Construction Award in 1989 and 1993, and L.A. County's Construction Safety Award for the past 3 years.

Project Schedule

Item	End of Week
Notification by client of award of project	0
Design:	
40% completion submittal	3
80% completion submittal	7
100% completion submittal	10
Final submittal for approval	12

4

FIGURE 15–3
(*continued*)

Begin construction	13
Grading operations complete	15
Utilities, and curb and gutter construction complete	17
Pavement construction complete	18
Landscaping, signing, and striping complete	19
Construction and inspection complete	20

Price

Design
150 hours @ $50/hr	$ 7,500	
5 sheets @ $30/sheet	150	
Reproduction and Administration	300	
Permits	2,000	
		$ 9,950

Construction
Labor
1800 hours @ $40/hr	$ 72,000	
Materials	50,000	
		122,000

Total Project Cost: $131,950

5

KEY CONCEPTS

- In the private sector, bids and proposals can keep companies in business or cause their demise.

- The presentation of your proposals indicates to the reader the thoroughness and accuracy of your work. Therefore, it is important that the development and organization of your proposals are complete, easy to understand, and professional-looking.

STUDENT ASSIGNMENT

1. Assume you are a designer and builder. Respond to an RFP from a college or university for one of the following projects. You may create purposes and contexts for facts and data needed for your response.
   a. On-campus student housing for 150 students.
   b. An Olympic-sized swimming pool.
   c. A centralized computer center.
   d. A technical research library.
   e. A faculty/staff cafeteria.

2. Assume you are the president of a manufacturing company. Respond to an RFP from the purchasing department of a college or university for one of the following:
   a. 200 student chairs with writing surfaces.
   b. A tension-testing machine for the strength of materials laboratory.
   c. 40 computer keyboards.
   d. 50,000 plastic book bags for use by the bookstore.
   e. 1000 felt-tip markers for board use.

3. Assume that you are the president of your student organization on campus. Submit an unsolicited proposal to the appropriate office or official on campus to obtain or accomplish one of the following:
   a. Receive approval for use of facilities and allocation of financial resources for a tutoring service by your organization.
   b. Add (or delete) a required class in your curriculum.
   c. Allow the community-at-large to use the university recreational facilities to promote goodwill. Use of the facilities would be by purchase of a yearly membership.
   d. Close (or open) a street to through traffic to facilitate pedestrian safety (or ease traffic congestion) on campus.

4. Assume you are a graduate student pursuing a doctoral degree in your field of specialization. Write a grant proposal to an appropriate organization (e.g., the National Science Foundation) to obtain the necessary funds to do research for your thesis. Select your own topic for this research. Assume that no specific format is required (although this is not generally true), and use the headings suggested in Figure 15–1.

CHAPTER 16

Specifications

Organizations use specifications to communicate technical requirements.

Clients and other purchasers of technical services (e.g., test laboratories and consultants); products (e.g., manufacturers); and projects (e.g., contractors) use specifications to communicate the technical requirements to providers of these services, products, and projects. Specifications control the quality of the services, products, and projects supplied by these providers. The term *clients* in this chapter includes all purchasers of services, products, and projects. The term *vendors* includes all providers of these services, products, and projects.

Because specifications are the basis for technical compliance with clients' requirements, all technical details must be clearly and unambiguously stated to prevent legal disputes.

PURPOSE AND AUDIENCE

Clients are responsible for writing specifications, whereas vendors are responsible for complying with them. However, vendors may be required to write procedural specifications in response to the clients' specifications. Other documents, similar to specifications in format, that are used in conjunction with specifications are discussed next:

1. *Contracts.* Contracts address the nontechnical agreements between two or more parties. They include such items as time for completion, terms of payment, and negotiation of disputes. A contract is made for a specific service, product, or project, and, for technical agreements, a contract ordinarily incorporates plans (drawings) and specifications.

2. *Codes.* Codes address the design parameters (see Chapter 5) for performance characteristics (e.g., the reasonable and intended temperature range for use of motor oil); the intended use of projects and products (e.g., sufficiently wide for wheelchair access); and safety requirements (e.g., fire retardancy of materials). They ensure uniformity in design criteria and are ordinarily written by a committee of industry professionals appointed by a professional association.

Safety codes, such as fire and building codes, ensure minimum standards for the safety of the users of products and projects. Compliance with safety codes is ordinarily required by local, state, or federal law, even when these codes are not specified in the contract.

Codes address generic classes of services, products, and projects rather than a specific service, product, or project. For codes other than safety codes, unless required by contract, compliance is voluntary.

3. *Standards.* Standards address the technical requirements (see Chapter 5) for the details of design and for methods of manufacture, construction, and measurement. Standards may be written by a committee of industry professionals appointed by a professional association to ensure uniformity (as in the construction of highways) or by an organization intending to reduce specifying repetitive details by its own professionals (as in the selection of chemical compositions of plastics for a company that uses plastics in the design of its products, or the measurement of operational characteristics of electronic components).

Organizations that mass produce products have pre-established standards of design and workmanship to determine the acceptance and rejection of their production items. When an organization claims to have high standards, it is referring to standards of design and workmanship that ensure satisfactory performance and quality for the consumer.

In summary, contracts and specifications specify the client's nontechnical and technical requirements to be complied with by vendors; codes specify the minimum requirements for design determined by industries and the law; and standards identify pre-established designs, methods, and tests that can be used to comply with codes and specifications.

Example

In the design of an office building:
- The <u>contract</u> may specify the penalty for late completion (e.g., "The contractor shall forfeit $500 per day for completion beyond the agreed upon completion date."). It may also specify insurance and bonding requirements and payment of subcontractors. It may incorporate the plans and specifications into the contract.
- The <u>specification</u> may specify the number of offices in the building (e.g., "There shall be 6 offices for each floor above the ground level."). It may also specify the materials and methods for construction, specific procedures for testing the electrical wiring, the

location of vending machines, and the requirements for recreational facilities and a library.

Applicable portions of codes, standards, reports, and other specifications for compliance by the vendor are cited in the body of the specification. These documents are listed in the "Applicable Documents" section of the specification (see Chapter 7).

- The codes may specify the minimum design unit weight to use for file cabinets (e.g., "Use a minimum weight of 200 lb for each three-drawer file cabinet."). The code may also specify the design equations to use for steel construction, the minimum ventilation requirements, and the minimum number and location of fire exits.

- The standards might include established designs of doors, window casings, and fans; methods of mixing cement and fastening joints; test procedures to demonstrate the adequacy of electric heaters; and minimum criteria for acceptance of elevator performance.

CONTENT

Specifications are often designated as design, test, bid, qualification, procedural, performance, or project specifications, and others, depending on their intended functions. Regardless of their designations, specifications are procedural specifications (procedures are specified rather than outcomes or performances); performance specifications (outcomes or performances are specified rather than procedures); performance specifications with procedural clauses within them; and procedural specifications with performance clauses within them.

Specifications almost always use block paragraphs and indent the text from the section numbers (commonly referred to as *outdenting*) for easy reference.

Procedural Specifications

Procedural specifications specify technical processes (e.g., arc welding) and laboratory procedures (e.g., determining the efficiency of an automobile engine). These specifications include the background, materials and equipment, and instructions or procedure (see Chapter 5, "Instructions and Procedures") that are required to understand and perform these processes and tests.

When a client uses a procedural specification, the vendor is responsible for complying only with the procedure in the specification. The client is responsible for the outcome.

Technical Processes   To obtain a quality product, clients often use procedural specifications to specify the procedure used by vendors such as certain manufacturing processes.

New and experimental materials and techniques frequently have unpredictable outcomes. Therefore, clients will use procedural specifications to encourage vendors to bid on products and projects based on vendors' abilities to perform these procedures without requiring them to guarantee an outcome.

Laboratory Procedures   Laboratory procedures are either of two types:

1. *Fragility procedures.* Fragility procedures, used primarily in research, determine the maximum capabilities of the tested items. For example, materials may be tested to failure, and pumps may be run to capacity. The results of these tests are usually for information that may be used in subsequent tests or design.

The procedures (experiments) that you perform in your laboratories at school are typically fragility procedures, and the explanations in your laboratory manuals to perform these procedures are procedural specifications to determine these maximum capacities.

2. *Qualification procedures.* Qualification procedures demonstrate that the tested items comply with the clients' specifications. For example, when the joint at the end of a beam is required by the client's specification to react 280 ft-lb of moment, the test is terminated when the applied moment reaches that value. It is of no consequence that the joint can react a considerably greater moment.

Because the performance of qualification procedures typically causes some damage to the tested items (e.g., flaring, or permanent elongation of bolt holes under load), the acceptable limits of the damage are included in the "Acceptance Criteria" section (see Chapter 5).

The following list of sections may be used as a guideline for procedural specifications (Figure 16–1):

1. *Introduction:* the purpose of the test and the source of authorization.

2. *Scope:* the parameters that are tested and any limitations on the applicability that form the basis of the test.

3. *Description of the tested structure:* a description of the test specimen.

4. *Test equipment and setup:* a description of the laboratory test equipment and setup. When the equipment is commonly used, a list is adequate and no description would be necessary.

5. *Test instructions or procedure:* the step-by-step laboratory instructions or procedure.

6. *Acceptance criteria:* the acceptable limits of damage for qualification tests.

7. *Notification of testing:* a statement of the manner in which the client will be notified to witness the tests.

8. *Documentation:* a statement of the written evidence that will be submitted to the client after performing the tests.

Figure 16–1
Sections of a Procedural
Specification

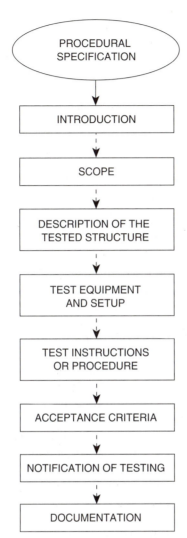

Sample Notification of Testing

The client will be notified of the time, date, and place of the testing 1 month ahead of time. This will give the client plenty of time to decide whether he or she wishes to witness the test.

Critique of Sample Notification of Testing

The information presented in this sample is adequate. However, two items deserve attention. "One month ahead of time" is a vague time reference and should be replaced with "1 month before the test." The second sentence, "This will give the client plenty of time to decide whether he or she wishes to witness the test" is an irrelevant cordiality and offers no new information. Therefore, it should be deleted.

SAMPLE ASSIGNMENT—PROCEDURAL SPECIFICATION

See Assignments 1 through 5 beginning on page 249 for the scenario. The facts of the sample assignment are as follows:

> Your company, Casesareus, manufactures Model 361 stereo casings that will use the Herehear label. Include, as a minimum, dropping, bumping, wear and abrasion, and liquid and chemical spills for a 7-year life as the structural and environmental loads.

Figures 16–2 (pp. 233–236) and 16–3 (pp. 237–242) illustrate two different samples of procedural specifications. A discussion of each follows.

Critique of Procedural Specification—Sample 1

Section 1 in Figure 16–2 correctly uses the phrase "will be manufactured." Note the important difference between *will* and *shall* in this context: "*Shall* be manufactured" would be incorrect here, because it implies that the client, rather than the manufacturer who is writing this specification, will be manufacturing these casings.

This sample appropriately uses the imperative tone in the test procedures. Each procedure requires between 8 and 11 steps. To simplify following the procedure, the procedure should be separated into sets of 4 or 5 steps, each set introduced with a subheading.

The acceptance criteria section appropriately states the minimum limits that are not acceptable, rather than the maximum limits that are acceptable.

The documentation section appropriately tells the client which reports will be submitted by the vendor and states the number of photographs that will be included in the report. Stating the number of photographs is common practice to prevent misunderstanding that would cause the reader to set up the test equipment again after the test is completed.

Critique of Procedural Specification—Sample 2

In Sample 2, shown in Figure 16–3, all section numbers are "outdented," as is typical, from the text of the sections.

In the scope, this sample includes, in list format, the conditions that will be tested. This makes these conditions more visible and easier for the reader to remember.

Sample 2 (see Figure 16–3), unlike Sample 1, uses sketches, which help the reader visualize the tested structure and the test equipment. (For space considerations, some of the sketches have been omitted in this text.) Referring to sketches and photographs in descriptions simplifies the text of these descriptions.

As in Sample 1, this sample appropriately uses the imperative tone in the test procedure and inappropriately uses many steps for each procedure, which should be separated into sets of four or five steps. (Text continues on p. 243)

Outdenting is common practice in specifications.

PROCEDURAL SPECIFICATION MODEL 361 STEREO CASINGS

1.0 Introduction

Casesareus has been manufacturing quality stereo casings since 1957. One thousand (1000) Model 361 casings will be manufactured using the Hearhere label, and will be qualified to the requirements specified in this document.

2.0 Scope

The test is intended to demonstrate that the casings can withstand the stresses imposed from all anticipated uses of the stereo equipment during shipment and a seven-year life in a home. Twenty casings will be chosen at random from the 1000 casings manufactured and tested for dropping, bumping, wear and abrasion and liquid/ chemical spills.

3.0 Description

3.1 Dimensions - The Model 361 stereo casing is 5 in. in height, 15 in. in length, and 8 in. in width.
3.2 Color - The Model 361 stereo casing is black.
3.3 Material - The Model 361 stereo casing is constructed of high-strength, low-alloy steel.

4.0 Test Apparatus and Setup

4.1 Bumping - The Model 361 stereo casing is tested for bump resistance on the BumpBump test machine manufactured by the Bumpsareus Corp. The BumpBump machine consists of a steel box twice the dimensions of the stereo casing, into which different material inserts may be placed to simulate the casing bumping against plastic and wood. The Model 361 is placed in the BumpBump box, and an elec-

The writer should be more specific, replacing "twice the dimensions" with numbers.

FIGURE 16–2
Procedural Specifications—Sample 1 (pp. 233–236)

Accompanying photographs of the tested structure and test apparatus would be helpful.

tronic shaker arm moves the box left and right, and forward and backward. All four corners and all four sides are tested for bump resistance.

4.2 Dropping - The Model 361 stereo casing is tested for the effects of dropping by the use of the Dropit machine, manufactured by the Dropsareus Corp. This machine mechanically raises the Model 361 stereo casing to a maximum height of 5 feet and releases its mechanical grip which allows the casing to freefall. The Dropit machine allows the Model 361 to be dropped from either a horizontal, vertical, or diagonal position.

4.3 Wear and Abrasion - The Model 361 is tested for wear and abrasion using the WearnTear machine manufactured by the Abuseit Corp. This machine vibrates Model 361 back and forth 3 ft at a maximum speed of 1 ft/sec to test for the effects that are due to friction. While the Model 361 vibrates across the 3-ft rough metal platform, mechanical arms apply pressure which a pair of scissors, a screwdriver, a pen, and a pocketknife. The Model 361 is tested with the WearnTear machine for 24 hours.

4.4 Liquid and chemical spills - The Model 361 is tested for resistance to liquid and chemical spills by pouring a predetermined amount of ammonia, nail polish remover (acetone), and coffee on all parts.

5.0 Test Procedure

5.1 Bumping.

5.1.1	Open the lid of the BumpBump machine and place the Model 361 into the steel box.
5.1.2	Secure the two buckles on the lid.
5.1.3	Center the two electronic shaker arms - one arm on the left and the other on the right side of the "shaker box."
5.1.4	Secure the nuts on the electronic shaker clamps with a crescent wrench.
5.1.5	Press the "on" button (red) and run the BumpBump machine for 5 hours.
5.1.6	Press the "off" button (green). Remove the Model 361.
5.1.7	Record the effects due to bumping.
5.1.8	Snap the plastic insert into the BumpBump machine's steel box.
5.1.9	Repeat Steps 5.1.1 through 5.1.7.

FIGURE 16–2
(*continued*)

5.1.10   Snap the wood insert into the BumpBump machine's steel box.

5.1.11   Repeat Steps 5.1.1 through 5.1.7.

5.2 Dropping.

5.2.1   Extend the Dropit machine's mechanical grip, which is marked "horizontal" (yellow).

5.2.2   Place the Model 361 onto the mechanical grips.

5.2.3   Push the "raise" button (red).

5.2.4   Select the height to raise the casing. Begin with a height of 1 ft and increase in 1-ft increments to 5 ft.

5.2.5   Push the "release" button (green).

5.2.6   Record the effects due to dropping.

5.2.7   Repeat Steps 5.2.2 through 5.2.6 using the vertical mechanical grips (blue).

5.2.8   Repeat Steps 5.2.2 through 5.2.6 using the diagonal mechanical grips (white).

5.3 Wear and abrasion.

5.3.1   Raise the mechanical arm to its vertical position.

5.3.2   Four mechanical grips attach to the arm, each grip with a different instrument attached to it.

5.3.3   Ensure that all instruments are securely fastened to the mechanical grips.

5.3.4   Center the Model 361 on the 3-ft rough metal platform.

5.3.5   Lower the mechanical arm to its horizontal position.

5.3.6   Push the "on" button (red).

5.3.7   After 24 hr, push the "off" button (green).

5.3.8   Raise the mechanical arm to its vertical position.

5.3.9   Record the effects due to wear and abrasion.

5.4 Liquid and Chemical spills.

WARNING: WEAR RUBBER SAFETY GLOVES AND PROTECTIVE EYE GOGGLES FOR THIS TEST.

5.4.1   Place the Model 361 in the plastic wash tub.

5.4.2   Pour 4 cups of ammonia over the Model 361, ensuring that all parts come into contact with the liquid.

5.3.2 is not an instruction and belongs in the test apparatus and setup section.

Note that the first entry in Section 5.4 is appropriately labeled as a warning and is not numbered as a procedural step.

5.4.3   Lift the Model 361 off the bottom surface of the tub, and then set it back down in the fluid.

5.4.4   Record the effects that are due to the chemical/liquid spills.

5.4.5   Remove the Model 361 from the plastic bin.

5.4.6   Dispose the liquid/chemical, and wash the plastic bin and the Model 361 with soap and water.

5.4.7   Repeat Steps 5.4.1 through 5.4.6 using nail polish remover.

5.4.8   Record the effects due to liquid/chemical spills.

6.0   Acceptance Criteria

6.1   The maximum deformation shall not be greater than 3/8 in. on any one side.

6.2   The maximum deformation due to the pressure of scissors, a screwdriver, a pen, and a pocketknife shall not be greater than 1/32 in. in depth or greater than 1 in. in length.

6.3   Wear of the Model 361 that is due to friction shall not be greater than 1/32 in.

6.4   No changes in color or material properties shall occur.

Color and material properties are unrelated criteria; color should be addressed in paragraph 6.4. Material properties should be separated into paragraph 6.5.

7.0   Notification of Testing

The client will be notified of the time, date, and place of the testing 2 weeks before testing.

8.0   Documentation

After the testing is completed, a test report will be sent to the client for approval. The report will contain a certificate of compliance with the test requirements and four photographs.

FIGURE 16–2
(*continued*)

PROCEDURAL SPECIFICATION
MODEL 361—STEREO CASING

1.0 Introduction

Casesareus, a leading manufacturer of stereo casings, will produce one thousand (1000) Model 361 stereo casings. The casings will be manufactured using the Hearhere label. This test specification for the Model 361 is to demonstrate compliance with the client's product specification.

2.0 Scope

Fifteen Model 361 stereo casings will be tested and evaluated for four loading and environmental conditions:

1. Dropping
2. Bumping
3. Wear and abrasions
4. Liquid and chemical spills

To simulate the weight, the electronic components will be installed before testing. The tests and evaluations will demonstrate that the stereo casing can resist the stresses experienced in shipment and a 7-year life.

3.0 Description of the Tested Structure (see Fig. 1)

The Model 361 stereo casing is made from 1/8-inch-thick, high-strength polyvinyl chloride (PVC). The casing is 28 inches long, 10 inches wide, and 12 inches high. It has slots in the front for the tuning, volume, balance, and equalizer controls. A hole, 4 inches long by 2 inches wide, in the front of the casing houses the LED station, volume, and signal displays. Another hole on the front of the casing, 6

FIGURE 16–3
Procedural Specifications—Sample 2 (pp. 237–242)

inches long by 4 inches wide, contains the cassette player/recorder. Two 6-inch-diameter speaker holes are in the front of the casing, one on each side. Cooling and sound vents are located in the rear and ends of the casing. Plastic posts on the inside are used to attach the electronic components to the casing. The casing is painted ferrous gray and, with the electronics installed, weighs 5.5 pounds.

4.0   Test Equipment

4.1   Dropping - The Freefall Dropper, built by Drops, Inc., simulates dropping the Model 361 stereo casing on wood, linoleum, and concrete surfaces. Two adjustable grippers hold the casing while it is mechanically raised to a maximum height of 8 feet (see Fig. 2). The grippers are released, which allows the casing to drop from the desired height.

4.2   Bumping - The Bump-O-Matic machine, built by Headbump, Inc., tests the Model 361 for resistance to bumping. The casing rests on the steel test surface which is surrounded by 8 panels (see Fig. 3) [omitted for space considerations]. Each steel panel is 24 inches long, 12 inches high, and 1/2 inch thick. Two columns of two impact arms on each panel are driven by springs with adjustable tension. These arms can be released individually or in any sequence with the other arms. Impact heads made of steel, wood, and plastic are screwed into the ends of the impact arms. The Model 361 can be placed on the test surface in any position and subjected to bumping by the impact heads.

4.3   Wear and abrasion - The Tearitup machine, built by P. L. Labs, tests the Model 361 stereo casings for wear and abrasion. The casing is clamped to the test surface and subjected to several moving instruments. A mechanical arm holds the test instrument and moves in a random direction (see Fig. 4) [omitted for space considerations]. The direction of the electric motor-driven arm is controlled by a joystick with a trigger for speed control. The arm adjusts up and down, which varies the pressure of the instrument on the casing.

4.4   Liquid and chemical spills - The Drencher machine, built by P. L. Labs, tests the resistance of the Model 361 stereo casings to liquid and chemical spills. The casing is clamped in place in the test tub,

FIGURE 16–3
(*continued*)

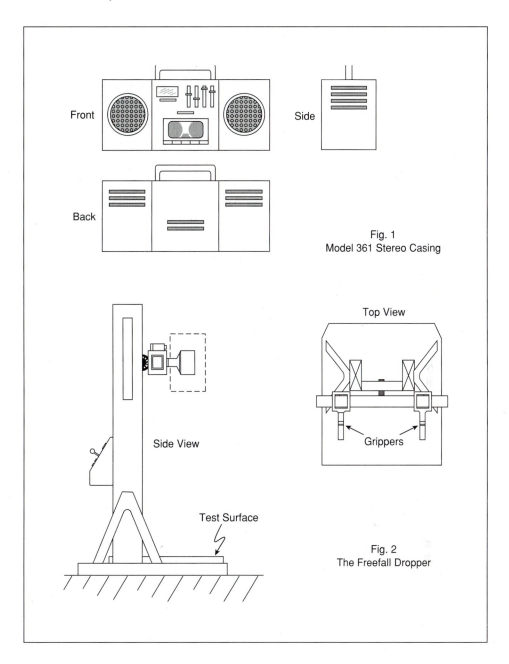

Front

Side

Back

Fig. 1
Model 361 Stereo Casing

Side View

Top View

Grippers

Test Surface

Fig. 2
The Freefall Dropper

and the reservoir is filled with a liquid or chemical, which is dropped on the casing (see Fig. 5) [omitted for space considerations].

5.0 Test Procedure

5.1 Dropping

5.1.1	Place the wood surface on the stand directly below the grippers.
5.1.2	Position the casing between the grippers.
5.1.3	Press the "close" button (yellow) to grasp the casing with the grippers.
5.1.4	Push the "raise/lower" lever (blue) up to raise the casing.
5.1.5	Raise the casing until it is 1 foot above the surface.
5.1.6	Press the "drop" button (red).
5.1.7	Inspect the casing and record the effects of the drop.
5.1.8	Repeat Steps 5.1.1 through 5.1.7 in 1-foot increments to a height of 8 feet.
5.1.9	Repeat Steps 5.1.1 through 5.1.8 using the linoleum surface.
5.1.10	Repeat Steps 5.1.1 through 5.1.8 using the concrete surface.

5.2 Bumping

5.2.1	Screw the wooden impact heads on the impact arms.
5.2.2	Place the casing on the test surface.
5.2.3	Press the release buttons (red) in any sequence or combination.
5.2.4	Simultaneously release all 32 impact heads.
5.2.5	Inspect the casing and record the effects from bumping.
5.2.6	Repeat Steps 5.2.2 through 5.2.5 using the steel impact heads.
5.2.7	Repeat Steps 5.2.2 through 5.2.5 using the plastic impact heads.

5.3 Wear and Abrasion

5.3.1	Clamp the Model 361 to the test surface.
5.3.2	Fasten a screwdriver to the mechanical arm (blue).

FIGURE 16–3
(*continued*)

5.3.3   Turn the pressure handle (yellow) to lower the arm.
5.3.4   Lower the arm until the gauge reads 1 pound.
5.3.5   Push the "on" button (green).
5.3.6   Grasp the joystick, and move the arm in all directions for 5 minutes.
5.3.7   Squeeze the trigger to increase the speed of the arm.
5.3.8   Adjust the pressure to 1/2-pound increments up to 5 pounds.
5.3.9   Push the "off" button (red).
5.3.10   Raise the arm from the casing.
5.3.11   Unclamp and inspect the casing for wear and abrasions.
5.3.12   Repeat Steps 5.3.1 through 5.3.11 using a ball point pen.
5.3.13   Repeat Steps 5.3.1 through 5.3.11 using a wood dowel.

5.4   Liquid and Chemical Spills
WARNING: WEAR PROTECTIVE GLASSES AND GLOVES FOR THIS TEST.
5.4.1   Place the Model 361 stereo casing in the test tub.
5.4.2   Fill the substance reservoir with soda.
5.4.3   Press the "drench" button (green).
5.4.4   Let the casing rest in the soda pool for 24 hours.
5.4.5   Remove the casing, and record the effects of spilling soda.
5.4.6   Drain the soda from the test tub.
5.4.7   Repeat Steps 5.4.1 through 5.4.6 using coffee.
5.4.8   Repeat Steps 5.4.1 through 5.4.6 using paint thinner.
5.4.9   Repeat Steps 5.4.1 through 5.4.6 using baking soda.

6.0   Acceptance Criteria

6.1   There shall be no cracks in the casing greater than 1/32 inch wide and 1/2 inch long.
6.2   There shall be no dents in the casing deeper than 1/8 inch.
6.3   There shall be no scratches or abrasions on the casing deeper than 1/16 inch.
6.4   There shall be no change in color of the casing.
6.5   There shall be no change in the size of the casing.

Criterion 6.1 is ambiguous. Does the writer mean "or 1/2 inch long"? As expressed, a crack 1/16 inch wide would be acceptable if it is less than or equal to 1/2 inch long.

7.0 Notification of Testing

Hearhere, Inc., will be notified 2 weeks before testing begins to witness the tests.

8.0 Documentation

Casesareus will submit a test report and the evaluation of the tested Model 361. We will also submit a certificate of compliance to certify that Model 361 complies with the performance specification.

FIGURE 16–3
(*continued*)

Step 5.4.1 is preceded by a warning (an instruction to prevent injury to the operator during normal operation) that is clearly labeled. However, the writer should clearly indicate to which step(s) the warning applies.

The acceptance criteria section, as in Sample 1, appropriately states the minimum limits that are not acceptable, rather than the maximum limits that are acceptable.

The documentation section tells the client what reports will be submitted by the vendor but, unlike Sample 1, neglects to state the number of photographs that will be included in the report.

Performance Specifications

Performance specifications specify physical characteristics and performance characteristics and requirements (outcomes and their limits, qualities, and quantities). Unlike procedural specifications, the procedures for accomplishing these outcomes are not specified.

Vendors are responsible for supplying products and projects that comply with their clients' performance requirements. Clients risk the possibility that vendors will use substandard techniques and materials for accomplishing these results.

It is common in performance specifications for clients to require vendors to demonstrate compliance of their products and projects with these specifications. To meet this requirement, vendors are usually required to submit, for approval, procedural qualification specifications to these clients.

The following sections may be used as a guideline for performance specifications (Figure 16–4):

1. *Introduction:* the general purpose and ownership of the product or project.

2. *Scope:* the work or project to be completed for compliance with the specification.

3. *Definitions:* a list of terms that have special meanings.

4. *Applicable documents:* a list of other documents that are required for compliance. The references to these documents in the list are cited in the body of the specification.

5. *Design parameters:* the physical and performance characteristics used for designing the product or project based on load and use.

6. *Technical requirements:* criteria for selection of materials, components, and procedures for design.

7. *Quality control:* the tests performed to demonstrate the physical and performance characteristics of the product or project. It also includes the limits of design and manufacturing defects for acceptance of the project or system.

Figure 16–4

Sections of a Performance
Specification

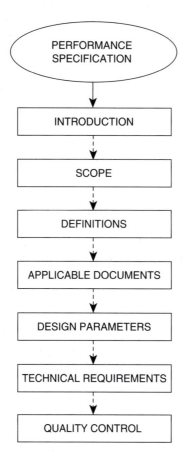

SAMPLE ASSIGNMENT—PERFORMANCE SPECIFICATION

See Student Assignment 6 on page 250. Figure 16–5 is an example of a performance specification.

Critique of Sample Performance Specification

The specification in Figure 16–5 is for bidding purposes; this student appropriately calls it a bid specification, even though it is also a performance specification. A list of definitions simplifies the explanations that are required within the text and eliminates misinterpretations.

Because no applicable documents section is included, no codes, standards, or other documents need to be complied with. Frequently, an applicable documents section is included and the applicable documents are cited in the body of the text.

Section 5.0 is titled "Design Requirements," because it encompasses the design parameters (the expectations of the users) as well as the technical requirements (selection of materials and components and methods of manufacturing).

Because the warranty in Section 7.0 is more appropriately included in a contract rather than a specification, it should be deleted.

Bid Specification

1.0 Introduction

1.1 This specification is for the design and construction of an oak television stand/video entertainment center. The owner, Jonathan Scott Roberson, will possess exclusive rights of ownership of the drawings, design, and constructed television stand. Upon award of the bid, this specification will become the legal document to specify the requirements of Jonathan Scott Roberson. A conceptual drawing is shown in Figure 1.

2.0 Scope

2.1 The successful bidder shall obtain the necessary materials, construct the stand, and finish all wood surfaces. Deviation from any item of this specification will require the approval of the client.

3.0 Definition

3.1 Client: Owner/client, Jonathan Scott Roberson.
3.2 Contractor: the successful bidder.
3.3 Videocassette: VHS format videocassette.
3.4 Hardware: Components of the television stand not made of wood (i.e., hinges, casters, drawer guides, and locks).
3.5 VEC: television stand/video entertainment center. The item to be constructed.

Figure 16–5
Sample Performance Specification (pp. 245–248)

FIGURE 1 — Video Entertainment Center

Note the use of "and" rather than "or" in "and other defects." "And" means all of these conditions shall be met.

4.0   Materials

4.1   Wood
    4.1.1   All wood shall be solid oak matched for grain pattern, color, and density.
    4.1.2   All wood shall be free of knots, blemishes, dry rot, and other defects.
    4.1.3   All wood shall be a minimum of 1/2 in. (0.5 in.) thick.

FIGURE 16–5
(*continued*)

4.1.4 Wood used on drawer interiors shall be a minimum of 1/4 in. (0.25 in.) thick.

4.1.5 All wood shall be sanded smooth with emery cloth no coarser than No. 200 grit or equivalent.

4.1.6 All wood surfaces shall not be stained or varnished.

4.1.7 Finishing of wood surfaces shall consist of one coat of Mr. Bear's Lemon and Almond Oil Furniture Polish applied to all wood surfaces.

4.2 Hardware

4.2.1 Exterior hardware shall be made of brass.

4.2.2 Console doors shall made from glass with a 25 percent smoke tint.

4.2.3 All hardware shall be inspected and approved by the owner.

5.0   Design Requirements

5.1 The VEC shall have envelope dimensions of 3 ft wide, 2 1/2 ft high, and 1 1/2 ft deep (3 ft × 2 ft 6 in. × 1 ft 6 in.).

5.2 The VEC shall be enclosed by wood on the top, bottom, back, left side, and right side.

5.3 A 9-inch drawer extending the full height and full depth of the VEC shall be constructed in the right side of the VEC.

5.3.1 This drawer shall include two interior tiers. The upper tier shall be positioned 1 ft above the lower tier.

5.3.2 Each tier of the drawer shall be enclosed by a 4-in. high piece of wood on the left and right sides.

5.4 A 1/2 in. thick wood partition shall be provided adjacent to the left wall of the drawer.

5.5 A shelf shall be provided 1 ft above the base in the area bounded on the left by the extreme left wall and on the right by the partition.

The shelf shall be mounted on metal tracks to enable it to slide forward in a manner similar to the drawer.

5.6   Two identical hinged glass doors shall be provided for access to the main storage area to the left of the drawer.

5.7   Magnetic strips shall be provided to secure the doors when closed.

5.8   A 1.0 in. high by 18.0 in. long hole shall be cut in the rear of the VEC 3.0 in. below the top of the unit.

5.9   Four casters, one at each corner, shall be provided to facilitate movement of the VEC over carpet and tile.

5.10 All connections shall consist of both a water-soluble resin-based glue and finishing nails. When feasible, nails shall be on hidden surfaces.

5.11 Nails on visible surfaces shall be sunk 1/16 in. below the wood surface and filled with wood putty to match the natural color of the wood.

6.0   Quality Control

6.1   The drawer shall deflect no more than 1/16 in. when 75 percent withdrawn.

6.2   The top of the VEC shall deflect no more than 1/16 in. when subjected to a vertical load of 250 pounds.

6.3   The shelf shall deflect no more than 1/16 in. when subjected to a vertical load of 40 pounds.

6.4   The drawer and shelf shall slide smoothly without binding.

7.0   Warranty

7.1   The contractor shall provide a 5-year warranty on parts, labor, and workmanship.

6.4 is vague. The writer should specify the pounds of force required to pull.

A warranty is part of the contract—not the specifications—and therefore should be deleted here.

FIGURE 16–5
(*continued*)

KEY CONCEPTS

- Specifications form the legal basis for establishing the requirements for technical work.
- Contractual requirements must be clearly and unambiguously stated to avoid misinterpretation.
- All requirements must be specified to control the quality of the service, product, or project.

STUDENT ASSIGNMENT

Procedural Specifications: Assignments 1 to 5

Read the following scenario, and write a procedural qualification specification:

Your company is a manufacturer of a Model 361 product for commercial suppliers. Your company has recently been awarded the contract to manufacture 1000 units of this product for a commercial supplier. The product will be marketed using the supplier's label.

Your client's performance specification for Model 361 product is included as part of this contract. Section 6.1 of this specification requires your company to demonstrate the capability of Model 361 to withstand the anticipated structural and environmental loads within its lifetime.

Section 8.0 of your client's performance specification requires you to submit, for approval, the procedural qualification specification for these tests and analyses that will be used to demonstrate compliance with Section 6.1.

Your assignment is to write the procedural qualification specification that will be used to demonstrate compliance of Model 361 with your client's performance specification. The relevant sections of your client's specification are as follows:

6.0   Structural and Environmental Qualification

6.1   Model 361 is to be qualified by test, or by a combination of test and analysis, to withstand the anticipated structural and environmental loads (e.g., temperature, chemical spills) imposed during shipment and its normal lifetime.

• • •

8.0   All quality control procedures and specifications are to be approved before manufacture.

Your client will not be familiar with the special-purpose hypothetical test equipment that you will use for performing the tests. Therefore, you will need to identify this equipment and briefly describe its mechanical operation.

1. Your company, Desksareus, manufactures Model 361 office desks, which will use the Writeright label. Include, as a minimum, dropping, bumping, wear and abrasion, and liquid and chemical spills as the structural and environmental loads for a 5-year life.

2. Your company, Framesareus, manufactures Model 361 backpacking frames, which will use the Sunisfun label. Include, as a minimum, dropping, abrasion, overstuffing, wear, chemical spills, rain, snow, humidity, cold, and heat as the structural and environmental loads for a 5-year life. Qualification is to include the effects of a backpack loaded with 60 pounds of simulated camping gear.

3. Your company, Shoesareus, manufactures Model 361 men's sportshoes, which will use the Runforfun label. Include, as a minimum, flexing, scuffing and scraping, impact, wear, and weathering as the structural and environmental loads for a 1-year life.

4. Your company, Skisareus, manufactures Model 361 racing skies, which will use the Paytoplay label. Include, as a minimum, the loads imposed by a rooftop carrier, storage in a garage, and use on the slopes including weathering, dropping, impact, flexing, and scraping as the structural and environmental loads for a 4-year life.

5. Your company, Campersareus, manufactures Model 361 campershells, which will use the Playtoday label. Include, as a minimum, induced vibration from a pickup truck, weathering, flexing, abrasion, and bumping as the structural and environmental loads for an 8-year life.

Performance Specifications

6. It has been several years since graduation, and you have been successful in your career. There has been something that you have wanted to own since high school. Money is no longer an obstacle to meet the needs of your sophisticated taste, and this item will have to be custom designed and built.

   Your assignment is to write a performance specification for manufacturers or contractors to bid on. On the award of the contract, the manufacturer or contractor must comply with the requirements of this specification. You may assume that the mechanical and electrical components that are required for this item are commercially available and that they will be assembled and packaged per the requirements of this specification.

   Suggestions for this item might be a bicycle, desk, guitar, or patio.

   A conceptual sketch may be helpful in communicating your ideas.

CHAPTER 17

Activity and Product Evaluations

Facts and data of the past are used in evaluations to determine the future.

Past events, completed procedures, and products frequently are evaluated to determine their value to participants and users. The results of these evaluations determine the desirability of continued participation in these events or continued use of these procedures and products. Typical evaluations might be the reason for the decline of sales of a product or service, the effectiveness of an etching process, and the performance of a new diesel engine.

PURPOSE AND AUDIENCE

Evaluations, based on past performances, help readers determine future courses of action. Only the relevant information should be investigated and evaluated. For example, evaluating the declining market sales of a product concentrates on assessing the marketing effort. The technical characteristics (i.e., functional and performance characteristics) of this product should be investigated and evaluated only to the extent that they affect the product marketability.

Readers of these evaluations are typically supervisors and clients but sometimes are the readers of newsletters and other publications (e.g., *Road and Track* magazine).

CONTENT

Evaluations investigate the component parts of their subjects to arrive at conclusions with regard to their wholes.

The results of evaluations include the advantages and disadvantages of the events, procedures, and products. Data and graphics should be included in the report to help readers understand the subject matter. The conclusions must be a direct result of the facts (objective truths) and data (numerical information) presented in the evaluation. A discussion of the alternatives and recommendations may be included, especially when the evaluation is written for a nonpeer audience that does not have the special knowledge or skills of the writer.

For example, the following components need to be considered when evaluating the effectiveness of a new structural test procedure:

- The reliability of the test data
- The availability of technical personnel with the expertise to perform this procedure
- The time required to perform the test
- The need for, and availability of, special equipment and tools
- The time required for equipment setup
- The need for disposable supplies

These components are analyzed using facts and data to arrive at a conclusion.

The cost of this procedure—which is of concern to an accountant—does not affect the efficacy of this procedure and therefore is not addressed.

Professionals reading this evaluation in a technical journal presumably know the effectiveness of the alternatives, and therefore they do not need to be discussed (although these alternatives should be identified in the introduction). Also, although these professionals need conclusions from which they can base decisions, they do not need a statement of recommendation.

Example

Exclude recommendations:	This test procedure should not be used for statically indeterminate structures.
Include a conclusion:	Using this procedure, the measured loads in the secondary members of statically indeterminate structures are accurate within 20 percent.

Professionals reading this conclusion should be sufficiently competent to determine future courses of action to obtain their desired accuracies.

However, interested nontechnical persons reading an evaluation in the technology section of a newsmagazine are concerned with the effectiveness of the alternatives and a recommendation from the informed writer. The cost can also be addressed in a newsmagazine.

The following sections should be included in activity and product evaluations. When these evaluations are only one or two pages long, they do not usually include headings (Figure 17–1):

1. *Introduction:* a statement of the item evaluated and the purpose of the evaluation.

FIGURE 17–1

Sections of an Activity and Product Evaluation

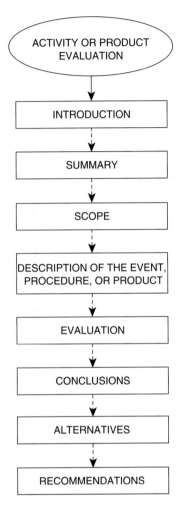

2. *Summary:* an abbreviated form of the report including the alternatives and recommendations, if any.

3. *Scope:* the concerns and exclusions of the study.

4. *Description of the event, procedure, or product:* a description to help the reader understand the event, procedure, or visualize the product and its functions.

5. *Evaluation:* the detailed analysis of the advantages and disadvantages of the event, procedure, or product.

6. *Conclusions:* what is concluded based on the evaluation.

7. *Alternatives:* a discussion of alternative events, procedures, or products that achieve the same result.

8. *Recommendations:* what is recommended based on the conclusions and alternatives.

An abstract may be included in evaluations found in technical journals.

Sample Scope from an Evaluation of a New Manufacturing Process

This new process is evaluated to determine the decrease in manufacturing costs of the Model 560-102 engine. The changes that have resulted from this process in the performance and reliability of this engine are not the subject of this evaluation.

Critique of Sample Scope

This section clearly identifies the concerns of the evaluation. Also, it identifies concerns that are excluded, although they may be relevant and significant in other contexts.

Sample Advantages (Included in the Evaluation Section)

The tent has an interior floor space of 44 sq ft, and 5 ft of headroom at the center. This is adequate space to comfortably sleep two campers with their gear or three campers. Because the floor is waterproof, I can set it up on grass and do not need to be concerned about water seeping through. The 36-in.-tall zippered door is adequate for comfortable entry and exit. The nylon pole construction is sturdy, and the tent does not sway more than 2 or 3 in. in the wind.

Because the tent has a large floor space and a waterproof tent fly, it can be used comfortably for playing games in inclement weather.

Critique of Sample Advantages

This subsection, and the next subsection, headed "Disadvantages" (not included in this text), are the student's evaluation of the tent. The first paragraph of this subsection discusses the technical features; the second paragraph discusses an alternate use of the tent. Appropriately, the advantages are expressed objectively and quantitatively (using numbers). For example, instead of writing "The tent is very large and has an abundance of floor space," the student includes the area of the tent in square feet and states the number of campers it can sleep, which allows the reader to judge its relative size.

Although this is an objective evaluation, in the third sentence, the student subjectively introduces the personal pronoun *I*. The purpose of the evaluation is to write about the tent, not the user's experience with the tent. Therefore, the third sentence should read "Because the floor is waterproof and water will not seep through, the tent can be set up on grass."

The student states five advantages within two paragraphs. These advantages would have more impact on the reader if they were placed in a list with each item preceded by a bullet. The third sentence in the first paragraph should eliminate the personal pronoun *I* and be rewritten, as discussed in the previous paragraph. Then, the wording in the list could be identical to the wording used in the sample, because the expression of the remaining sentences is parallel (even though two of the sentences have subordinate clauses beginning with the word *because*).

SAMPLE ASSIGNMENT—PRODUCT EVALUATION

From your perspective as a college student, write an evaluation for high school seniors concerning the effectiveness of your desk as a place of study.

Figures 17–2 and 17–3 illustrate two different samples of this type of product evaluation.

Critique of Product Evaluation—Sample 1

The introduction in Figure 17–2 appropriately discusses the purpose of this report ("so that other students . . . might benefit").

For reports with a limited scope, it is acceptable to include this scope in the introduction. In this sample, the scope is discussed in the introduction with the phrase, "discusses the advantages and disadvantages."

In any formal report, use of the subjective first person *I* should not be included. Rather, the objective voice should be used. For example, the opening sentence in the introduction could read "A college student has used a card table as a work space."

A summary is not included, which would help a reader not interested in reading the entire report determine the suitability of this desk for the possible application.

The description helps the reader visualize the desk. However, a sketch of the desk would be helpful.

This sample clearly separates the advantages from disadvantages, which simplifies the readability of the conclusion. The disadvantages are discussed before the advantages. Then, this report concludes that the desk is suitable for its intended purpose. This conclusion is appropriately based on the analysis in the advantages section, immediately preceding this conclusion. The conclusion appropriately includes details that justify the writer's opinion.

This sample does not address alternatives that may help readers determine if this desk is the best possible solution for their purposes.

Use of the first person may be appropriate when writing for a nontechnical peer group such as students. However, the objective voice is ordinarily used for formal reports. For example, this evaluation could begin, "A college student has used a card table. . . ."

Note the appropriate use of the present tense in the introduction: "This evaluation discussed. . . ."

Desk Evaluation

Using a Card Table as a Student Work Space

Introduction

I have used a card table as a work space for the past four quarters while attending classes at this university. There are disadvantages to performing school assignments on a card table, but I have found it to be a viable alternative to the traditional school desk. This evaluation discusses the advantages and disadvantages so that other students on limited budgets and in temporary living situations might benefit.

Description

The card table that I use as a desk was found in a garage. The top is 29 1/2 in. square and consists of a rusty 1 3/4-in.-wide metal frame with two 3/4-in.-wide diagonal cross members which support a cardboard top surface that sags. At one time, the top surface was covered with Contact paper which since has torn and peeled away at the corners to reveal the worn and weathered original surface and the rusty frame. The folding steel legs are 26 in.long and have a patina of rust from being used outdoors. The legs are connected to the frame by pins and may be folded so that they are contained within the frame of the top. When unfolded, the legs are locked into place by 1/2-in.-wide and 3-in.-long steel braces that are pinned to the legs at one end and attached to the top frame at the other end with pins which slide in a groove. The table weighs 15 lb.

Disadvantages

The primary disadvantage to using the card table as a work space is that it is too narrow to support the portable drafting board I use for design and writing. This drawing board is 36 in. wide and has a 4-in.-long folding leg at each of the rear corners to support it at the proper angle. Because the table is only 29 1/2 in. wide, I have placed a 40-in.-wide board between the top of

1

FIGURE 17–2
Product Evaluation—Sample 1 (pp. 256–258)

the table and the drawing board legs to support the rear of the board. The front of the drafting board rests on the frame of the table, straddles the sagging top, and requires a magazine shim on the left side to stabilize it. Due to this arrangement only a 4-in. strip of level surface is along the front of the tabletop. Because of the sag of the tabletop, pencils and other tools tend to roll under the drawing board.

Another disadvantage is that the card table has no drawers or shelves. Because the drafting board covers the entire top except for a 4-in. strip, there is no room to store books or materials.

Elevation of the tabletop is another problem. Because the legs are not adjustable, the working surface of the table is fixed at 27 3/4 in., which is too low to be comfortable for a tall person. Even if an inclined surface is employed to correct this, as described, clearance between the seat of a chair and the bottom of the card table may be insufficient for a person with large thighs to sit comfortably.

Last, card tables, especially dilapidated ones, are not sufficiently stable to support personal computers. Because the leg brace dimensions are too small (see Description) to allow for leg room, a 5-lb lateral force directed against the edge of the tabletop will move it 2 in. This much movement can be hazardous to delicate electronic equipment.

Advantages

The obvious advantage to using a card table as a desk is the low cost. Most students can probably find one in their family's garage or attic, or they can buy a new one for a fraction of the cost of a used desk. A new card table at Price Club ($25.00) costs one-fifth the price of a used desk at the Salvation Army Store ($125.00). This savings can be important when considering the high cost of books and living expenses for students.

Another important advantage is the ease of moving. Because students change their places of residence on an average of 2.4 times during each calendar year, having furniture that can fold into a compact space eliminates the expense and inconvenience of renting or borrowing a truck or trailer each time they move. The folded table (volume = 0.85 cu ft) can be easily transported in a small car. A nonfolding desk of the same dimensions

2

The writer might include other advantages such as versatility and potential for resale upon graduation.

as the unfolded table (volume = 14.11 cu ft) would require the use of a van or truck. Loading a folded card table rather than a bulkier and heavier desk also reduces the risk of damage to the vehicle and the furniture and injury to the student.

Conclusion

After considerations of expense and ease of movement, a card table may be a practical alternative to a desk when certain conditions are met. Its top surface should be at a comfortable height for working. This is important because the legs cannot be adjusted. It should not be used for loads greater than it is designed for, which ordinarily excludes placing electronic equipment on it. Storage space for books and other materials must be provided. However, some of the characteristics I have listed as disadvantages may, in certain situations, become advantages. For example, a small card table may fit into a small room, and a person who needs more leg room may prefer no drawers.

3

FIGURE 17–2
(*continued*)

Critique of Product Evaluation—Sample 2

Similar to Sample 1 (see Figure 17–2), a sketch included with the description of the desk in Figure 17–3 would be helpful.

Unlike Sample 1, Sample 2 includes a separate scope section. Although the information in this section is limited and could be included in the introduction, this method is also acceptable.

This sample, similar to Sample 1, also advantageously separates the evaluation into advantages and disadvantages. However, the many advantages are enumerated in a list to emphasize each point and to facilitate remembering by the reader. However, numbering the items in the list implies a sequence of operation; instead, bullets, or asterisks should be used.

Unlike Sample 1, Sample 2 discusses the advantages before the disadvantages and then also concludes that the desk is suitable for its intended purpose. This sample requires readers to think about the advantages before the disadvantages and then bases its conclusion on the advantages, a former analysis. This may confuse some readers.

This sample includes a recommendation to improve the utility of the desk based on the discussion in the disadvantages section. Based on the special experience of the writer, this recommendation is intended to alert readers of a simple feature to consider when choosing a desk.

This sample, unlike Sample 1, appropriately addresses alternatives (even though it concludes that none are more efficient).

The conclusion includes a subjective statement without the details that could help convince the reader of the report's conclusion.

Desk Evaluation:
A Study of H-H Engineers' Currently Available Drafting Table

Introduction

A person's work desk should allow that person to conveniently and effectively complete a specific task. The desk that is the subject of this evaluation is the drafting table at which I spend up to 8 hours a day drafting, designing, and writing.

Description

A hard vinyl sheet covers the drawing surface of the 31.5-in. × 60.0-in. × 1.25-in. wooden desk top, which is supported by four 2.3-in. × 2.3-in. × 35.0-in. wooden legs, one in each corner. Connecting the front and back legs, two 1.25-in. × 2.5-in. × 25.0-in. struts, one on the left side and one on the right side, lie horizontally, 6.0 in. above the ground. Connecting these two struts are two 1.25-in. × 2.5-in. × 53.0-in. members lying horizontally 6.0 in. apart under the center of the table. The front edge of the tabletop is hinged to a 1.0-in. × 2.0-in. × 51.5-in. member, which extends between the two front legs. Two additional 7.5-in. × 1.0-in. × 25-in. members, one on the left side and one on the right side, extend between the top of the front and back legs and act as a rest for the tabletop when it is not in its tilted position. Located on these two members are the setscrews which pinch the rods that are connected to the underside of the tabletop. These rods can slide up or down, allowing variable tilt of the drawing surface. Two drawers fit beneath the tabletop: one is 27.0 in. long, 38.0 in. wide, and 1.75 in. deep, and the other is 27.0 in. long, 10.5 in. wide, and 3.75 in. deep. A sectional tray (14.0 in. × 10.5 in. × 1.0 in.) slides along a track located halfway up the inside of the deeper drawer and provides two levels of storage.

1

FIGURE 17–3
Product Evaluation—Sample 2 (pp. 260–262)

Rather than numbers, bullets should be used in this list, because this list does not show a chronological sequence.

Scope

The desk is evaluated for its use by a draftsperson only. Its adaptability for writing reports and performing clerical tasks is not addressed.

Advantages

This type of desk offers many advantages for drafting:
1. Its structure and material provide a sturdy work station.
2. A drafting arm and a light can be clamped securely on an edge of the table.
3. The smooth drawing surface is appropriate for a person to draw and write precisely.
4. The tabletop is sufficiently large to have both drawing and drafting materials on it at the same time.
5. The drafting surface can tilt sufficiently to allow access to all areas of the drawing, without the drafter leaning on the drawing.
6. The height of the table allows the drafter to either sit or stand.
7. The flat drawer is sufficiently large to store large drawings.
8. The deeper drawer is convenient for storing miscellaneous drafting materials, timesheets, or small personal items.
9. The two support boards that connect the left side to the right side of the table provide a foot rest for the drafter.

Disadvantages

The disadvantage with this desk is that when the drawing surface is in its tilted position, pens, papers, and miscellaneous drafting equipment easily slide from the table, which disrupts the drafter's work.

Recommendations and Alternatives

To improve work efficiency, a ledge should be added to the front edge of the tabletop to prevent drafting items from sliding when the desk is tilted.

2

When the desk is not in a tilted position, the ledge could be conveniently folded out of the way.

There are no alternatives that are more efficient than using this desk for drafting purposes.

Conclusion

The advantages of the drafting table far outweigh the disadvantages. By providing a ledge as recommended, this desk can be an efficient work space for the drafter.

3

FIGURE 17–3
(*continued*)

KEY CONCEPTS

- Only facts and data used in your evaluation should be included in your report.
- Your conclusions must be justified by the facts and data presented in the evaluation.
- Graphics can be helpful for the reader to understand your evaluation.

STUDENT ASSIGNMENT

1. From your perspective as a college student, write an evaluation for other college students and peers on one of the following:
   a. The success of a recent fund-raiser by your student professional organization.
   b. The importance to your career of the general education class you most recently completed.
   c. The significance of a part-time or summer job in your field of study.
   d. The importance of your most difficult class.
   e. The efficiency of your route of travel to school.
   f. Your method of studying for final exams.
   g. The importance of laboratory classes in your educational experience.

2. From your perspective as a college student, write an evaluation for high school seniors on one of the following:
   a. The efficiency of the registration process at your college or university.
   b. The suitability of your living arrangement (apartment, dorm, etc.).
   c. Term papers or laboratory reports as an educational tool.
   d. The intellectual benefits of your major field of study.
   e. The cost-effectiveness of your mode of transportation.
   f. The convenience of the location of your residence.
   g. The effectiveness of your running shoes or other sports equipment.

Feasibility Reports

Potential courses of action are described in feasibility reports.

Feasibility studies analyze the technical and economic practicalities of potential courses of action. The results of these studies are reported in feasibility reports.

FEASIBILITY REPORTS COMPARED WITH ACTIVITY AND PRODUCT EVALUATIONS

Feasibility reports are similar to activity and product evaluations because, similar to activity and product evaluations, they discuss activities and products. However, feasibility reports discuss proposed activities and products, whereas activity and product evaluations discuss completed activities and products already in use. Therefore, feasibility reports use projected, rather than substantiated facts and data. Also, feasibility reports may discuss multiple proposed activities and products, whereas activity and product evaluations discuss only one completed activity or product.

PURPOSE AND AUDIENCE

Feasibility reports recommend courses of action based on technical and economic analyses. They are written in response to requests from management and clients in the following circumstances:

- To evaluate the potential for success of a proposed activity or product, such as a new marketing strategy, the replacement of an obsolete tool to improve production, or improved performance that is due to the purchase of a new product. The proposed activity or product is hoped to be an improvement over the present satisfactory activity or product.

Taking no action is the only alternative to the single proposed activity or product.

- To compare proposed activities or products to determine the best of multiple alternatives, such as the site location for a new facility or the purchase of a new pump. When the present activity is satisfactory, taking no action is an alternative. However, when the present activity is unsatisfactory, one of the proposed activities must be selected.

In an economic feasibility analysis, when:

- The present activity or product is satisfactory, and therefore, taking no action is an alterative to the proposed activity or product, for a proposal to be a satisfactory alternative, either
  - The benefits must exceed the costs (commonly used in private industry).

  or

  - The benefits divided by the costs must exceed 1.0 (commonly used by government agencies).

- The present activity or product is unsatisfactory, and therefore taking no action is not an alternative (e.g., action must be taken either by necessity of situation, or contractual obligation), the most satisfactory alternative then is selected based on either
  - The greatest benefits minus costs, even when the costs exceed the benefits for all alternatives (commonly used in private industry).

  or

  - The highest ratio of benefit divided by cost, even when this ratio does not exceed 1.0 for all alternatives (commonly used by government agencies).

A technical feasibility analysis is typically the result of significant research and development (see Chapter 23, "Research Reports"). The criteria for acceptance of an activity (e.g., an improved chemical process) or a product (e.g., a larger carburetor) are determined by comparing the increased benefits (e.g., increased acceleration using a proposed carburetor) with the increased cost in dollars plus the "disbenefits" (disadvantages) (e.g., increased fuel consumption and maintenance). Because the realm of possible technical benefits for all activities and products includes all possible technologies, and the desired magnitude of these technical benefits is an intangible (i.e., subject to one's emotions), a discussion of technical feasibility is beyond the scope of this text and not discussed further.

CONTENT

Feasibility reports typically include the following sections (Figure 18–1):

FIGURE 18–1

Sections of a Feasibility Report

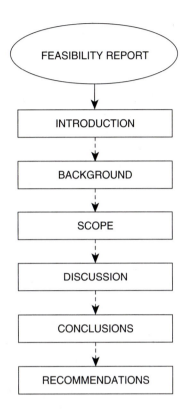

1. *Introduction:* a statement of the problem. Include the subject and purpose of the study and its authorization.

2. *Background:* the changes that occurred to create the necessity for this study.

3. *Scope:* the proposed alternatives to the present activity or project that are the basis of the study.

   When multiple proposed activities or products are compared, it should be clearly stated when the present activity or product (or its absence if no activity or product presently exists) is satisfactory. When only one alternative is proposed, the present activity or product (or its absence) is deemed to be satisfactory.

   Include the basis for selecting the proposed activities or products (e.g., data from scale model tests that are deemed to be sufficiently accurate) and conditions that limited the selection of proposed activities or products (e.g., the proposed activity must not require an inventory of spare parts, or the proposed product must not cost more than $2000/yr for operating expenses).

4. *Discussion:* the analysis of the proposed activities or products.

Include the information used to form the conclusions and recommendations of the study in this section.

Use facts, data, and calculations as the basis of your analysis. Present the information in a sequence that readers can easily comprehend.

For multiple proposed activities, analyze each activity individually before comparing them. Supplement your analyses with tables, charts, and graphs to demonstrate these comparisons.

Clearly explain the methodology that is used to justify the conclusions and recommendations.

5. *Conclusions:* the natural results from the information presented in the discussion. This section is the link between the discussion and the recommendations.

6. *Recommendations:* the course of action determined as a result of this study. The first recommendation discussed is the course of action that is the most advantageous to resolve the problem.

When a proposed course of action is recommended over the present course of action, it is appropriate to discuss other concerns, not the subject of the analysis. For example, for each recommended course of action, discuss the method, time required, internal capability of the organization, and availability of resources.

Also, related effects can be discussed such as the impact on other departments, other components, and the environment.

When the present course of action remains the preferred course of action, alternative courses of action may be suggested for further study.

Although the information in this section is a direct result of the preceding conclusions, this section may be placed after the introduction because of its usefulness.

Readers of feasibility reports may have nontechnical backgrounds. Therefore, write the report without jargon.

Sample Discussion
(Calculations Shown for a Rental Truck)

	Expense	Income
Purchase price	$18,000	
Maintenance, 5 yr @ $350/yr	1,750	
Tires, 5 yr @ $150/yr	750	
Overhaul @ 4 yr	1,500	
Rental, 5 yr @ $4,500/yr		$22,500
Resale @ 5 yr		5,000
TOTAL	**$21,750**	**$27,500**

Benefits minus costs = $27,500 − $21,750 = $5,750 > $0 OK

Critique of Sample Discussion

The calculations are clearly presented and easy to understand. Recurring items include the frequency and amount. The two columns, expense and income, organize the entries that must be added together. The totals, in bold numbers, are easily visible. The benefits minus the costs are calculated, shown to be greater than 0, and declared to be satisfactory. Because the benefits minus the costs are calculated, it is presumed that study is for a rental truck in private industry. Notice that if the benefits/cost ratio were calculated, as is typical for government agencies, this ratio would be 1.26 and would also be deemed to be satisfactory.

Because the largest number in each column exceeds four digits and therefore requires a comma, appropriately, all numbers in each column that include four or more digits also include a comma (remember that in technical writing, four-digit numbers do not ordinarily include a comma). Also notice that the commas align.

Did you notice the arithmetic error? The expense column should total $22,000. These errors are not uncommon for young professionals (arithmetic errors occur in many student reports). These errors, which easily can be corrected, create doubt, not only concerning the calculations, but also concerning the reliability of the assumptions, theory, conclusions, recommendations, and any other section of a report that requires critical thinking.

Sample Recommendations

I recommend that the drafter reject the job offer for more money because of the travel required. The additional income would more than offset the additional travel expenses. However, her present income is adequate, and the time spent traveling could be spent with her family. If she believes that she should be earning more money, she should continue to pursue new employment closer to home.

Critique of Sample Recommendations

The first sentence rejects the proposed activity with justification, as a result of this study. The following two sentences acknowledge the benefit of the proposed activity to the reader but justify its rejection.

The last sentence proposes an alternative course of action to achieve the same result. This sentence should begin a new paragraph because of its importance.

This section is easy to understand and, in spite of the negative recommendation, has a positive effect on the reader because of the alternative recommendation.

SAMPLE ASSIGNMENT—FEASIBILITY REPORT

Create purposes and contexts for any factual information and data needed for the following assignment.

A fellow classmate is considering the purchase of a used car to take a part-time, career-related job off campus. Your classmate is self-supporting with the aid of a 15-hr/wk job in the bookstore, which pays $6/hr; summer employment in the family business, which pays $350/wk; and a $2500/yr student loan. This job pays $12/hr but requires your class-mate to work 4 hours every afternoon.

Because of the additional hours required to be spent at work and the desire to maintain an A− average, your classmate will need to carry a reduced study load. This will delay graduation by only one semester (quarter) if the student attends summer school. Assume that the car costs $3000. Payments will be $87/mo for 3 years with $500 down. Insurance and maintenance will be $150/mo.
Write a feasibility report for this classmate to recommend a course of action.

Figures 18–2 and 18–3 are two examples of responses to this assignment.

Critique of Feasibility Report—Sample 1

In Sample 1 (Figure 18–2), the scope, which is necessary to alert the reader of what is discussed, is inappropriately not included. Also, this sample does not state the data given to perform the analysis. This information is crucial and needs to be included.

The advantages and disadvantages of career-related jobs and used cars are discussed in unnecessary detail in the discussion. Although details should be included, including unnecessary details implies a lack of basic knowledge of the reader.

In spite of the arithmetic error in the calculations, the structure makes the cal-culations easy to follow. Each course of action is labeled, and the totals are included.

However, the data for these calculations are not stated in the proposal or the analysis. Also, the report needs to show how each number in the columns is calcu-lated. For example, the bookstore salary should read:

Bookstore salary   15 hr/wk $\times$ 30 wk/yr $\times$ $6/hr = $2700

In the conclusions and recommendations, Sample 1 summarizes the advantages of the proposal. In spite of the heading, no factual conclusions are stated as a result of the analysis, and the recommendation is merely implied in the closing of the last sentence where the author states "it will help this student." The recom-mendation should be clearly stated such as "John should take this part-time, career-related job to give him an advantage in finding full-time employment in civil engineering when he graduates."

The recommendations are frequently written as a list for easy interpretation by readers.

FEASIBILITY REPORT: A CAREER-RELATED JOB

In discussing the salary, the writer should replace "will be double" with numbers.

Introduction

In order to take a part-time civil engineering job, Janice Alk must buy a used car to travel to it. Although the new job salary will be double that of her present job, she has to consider many additional problems which include staying an additional quarter at school to maintain her A– average, additional tuition for that quarter, travelling expenses, and insurance and payments for the car. However, it may be desirable to take the career-related job because the experience may help Janice to find employment upon graduation.

Discussion

Career-related jobs. A career-related job can help a student in many ways. It offers an opportunity to determine career interests, gain experience, and make personal contacts. Career-related jobs also give valuable insights into the function and politics of a civil engineering company. The student can learn first-hand about the civil engineering company and the specific job functions she or he can expect to find upon graduation. Also, a career-related job can provide practical experience for a student as she or he earns money to pay for school. A career-related job gives the student an advantage over the hundreds of other civil engineering graduates seeking employment. Also, the student can hope to receive a job offer from the firm at which she or he works.

The disadvantages for taking this job are the delaying of graduation and the expenses for buying, insuring, and maintaining the used car.

Used car. Although it is very convenient to own a car, a used car can create stress, because a used car may develop problems such as engine and transmission trouble. The buyer must worry about unknown wear and tear caused by the former owner or owners and future expenses for mechanical maintenance. The used car may have defective features such as worn safety belts, oil and radiator leaks, noise in the engine, and ignition problems. Because of wear, the steering response in an older car may not indicate to the driver how his or her tires are gripping the road and may cause an accident. The older car may be less fuel-efficient and slow in accelerating. Braking efficiency may be diminished in the used car.

1

FIGURE 18–2
Feasibility Report—Sample 1 (pp. 271–272)

Yearly Financial Analysis

Present Job

Bookstore salary	$ 2,700
Summer job	5,250
Student loan	2,500
Total annual income	$10,450

Career-Related Job

Salary	$9,600
Student loan	2,500
Down payment for car	− 500
Insurance and maintenance	−1,800
Annual payment for car	−1,044
Additional tuition fee	− 278
Total annual income	$8,725

Conclusions and Recommendations

Recruiters at major civil engineering companies receive hundreds of resumes annually; it is impossible for them to interview every person looking for an entry-level position. Therefore, recruiters frequently skim through resumes to study the applicants' relevant experience. Having a career-related job will give the student an advantage over other competitive applicants. Although taking a career-related job will require travel time, car maintenance and expenses, and provide less spending money, it will help this student to find a job when she graduates.

2

This section should show the calculations used to get these numbers, for example, 15 hr/wk × $6/hr × 30 wk = $2,700.

Note the arithmetic error for total annual income. The total should be $8,478.

FIGURE 18–2
(*continued*)

Critique of Feasibility Report—Sample 2

Sample 2 (Figure 18–3) includes a statement of the problem instead of an introduction, an acceptable substitution. The problem and the proposed solution are discussed in this section; the history of the proposal is discussed in the background; and the assumptions of the analysis are discussed in the scope.

Although the proposed solution is usually discussed in the scope, its placement with the statement of the problem is satisfactory.

In the background, this sample briefly discusses the advantages of the proposed course of action with relationship to the history of the proposal without including the unnecessary details of owning a used car. This appropriately assumes the reader has the basic understanding of the advantages and disadvantages of the proposal and merely needs to be reminded.

The calculations in the discussion have a clerical error ($11,420.00 should read $5,950, but the total $11,420.00 is correct), and a data error (payments of the car are $87.00/mo rather than $87.50/mo).

In the discussion, each set of calculations should have a heading. The results are further calculated in the paragraph following the tabulated calculations. This information is difficult to understand and should be included in a table.

The opening statement in the recommendation is the recommended course of action as a result of this study. The benefits of taking this course of action are then briefly reiterated. The closing sentence states what the student must do to implement this proposal. This section is clearly written and easy to understand.

FEASIBILITY REPORT
FOR WORKING STUDENT

Statement of the Problem

A university student is considering the purchase of a used car so that he can take a part-time, career-related job off campus. He is self-supporting with the aid of a 15 hr/wk job in the university bookstore, which pays $6/hr.; summer employment in his family's business that pays $350/wk.; and a $2,500/yr student loan. The new job pays $12/hr for 4 hours in the afternoon, Monday through Friday. The additional hours at work would require him to reduce his study load and go to summer school. This would delay his graduation by one quarter. The car would cost $3,000 with a down payment of $500; the monthly payments are $87.00. Insurance, fuel, and maintenance costs would total $150/mo. Is this course of action feasible?

Background

Choosing between school and work is a common dilemma for university students. As the percentage of high school graduates who enter colleges or universities has risen, so has the percentage of students from mid- to lower-income families. These students cannot rely on their parents to support them, and many work to earn their living and educational expenses. Because a greater number of graduates enter the job market every year, employers are looking for prospective employees who have work experience as well as a diploma. This has given students the additional incentive to work part-time in a job related to their field.

The scope section should describe for whom or under what conditions the report is applicable, not the assumptions of the report.

Scope

For lack of statistical date concerning the attitudes of employers regarding the importance of grades, of timeliness of graduation, and of career related work, this report assumes equal weight for all these factors. It assumes that the student plans to accept full-time work in his chosen field immediately on graduation. It further assumes that the student will attend school for 3 more years and will receive no raise in salary or obtain additional income. Also, it assumes that the student will incur no additional debts.

1

FIGURE 18–3
Feasibility Report—Sample 2 (pp. 274–276)

Note the clerical error: The summer job income should be $5,950.00 not $11,420.00.

Discussion

Because the car payments are spread over 3 years, the average net income per year is calculated below for each alternative.

If the student keeps his job in the bookstore and works for his family in the summer, his net income would be as follows:

Bookstore	$6/hr $\times$ 15 hr/wk $\times$ 33 wk/yr =	$2,970.00
Summer job	$350/wk $\times$ 17 wk =	11,420.00
<u>Student loan</u>	$2,500/yr =	<u>2,500.00</u>
TOTAL		**$11,420.00**

If the student buys the car and accepts the new part-time job his net earnings would be:

Income from:		
New job	$12/hr $\times$ 20 hr/wk $\times$ 50 wk/yr =	$12,000.00
Student loan	$2,500/hr =	<u>2,500.00</u>
Subtotal		**$ 14,500.00**
Less cost of car:		
Payments	$87.50/mo $\times$ 12 mo/yr =	$1,050.00
Down payment	$500/3 yr =	166.67
Maintenance	$150/mo $\times$ 12 mo/yr =	<u>1,800.00</u>
Subtotal		<u>**$3,016.67**</u>
TOTAL		**$11,483.33/yr**

Note the clerical error: The payments should be $87.00/mo not $87.50/mo. The annual payment should be $1,044.00.

The data in this paragraph are difficult to follow and would be more effectively presented in a table.

Although the student will net slightly more over the years if he accepts the new job, he will also work 3 months longer at his part-time job. Assuming the full-time entry-level position pays the same $12/hr, the student will lose $2,640.00 over the 3 months rather than starting his full-time career as originally intended. His net earnings for the 39-month period would thus be $16,700.00 if he keeps his present job and $14,123.33 if he takes the new one - a difference of $2,576.67. The student would have to spend much of this on a car after graduation to commute to work.

Recommendation

The student should choose the part-time job that best prepares him for his future work. Although keeping his campus job may facilitate graduating sooner and beginning his career, it will not provide him the experience in his field of work. This experience will give him a better bargaining situation for his starting position and salary. In addition to this, owning a car will provide him with the opportunity to visit prospective employers and become more involved with the outside community. Therefore, the student should buy the used car and accept the career-related off-campus job.

3

FIGURE 18–3
(*continued*)

KEY CONCEPTS

- The possible recommendations are limited to the proposed alternatives. Consider all options before you begin your analysis.
- The recommendations are the reason for the feasibility report. These recommendations include the judgment of the writer and should be carefully reviewed to provide the best solution to the problem.

STUDENT ASSIGNMENT

Create purposes and contexts for any factual information and data needed for one of the following:

Single Proposed Course of Action

1. Your cousin is completing her junior year at a private university in the same town where she lives. Tuition, $10,000/yr, is paid for by your aunt and uncle; living expenses, $5,000/yr, are paid for by your cousin with a part-time job during the school year and summer employment. Because your uncle has had a financial setback in his business, he is considering sending your cousin to the state university that is located 75 miles away for her senior year. Tuition at the state university is only $1500/yr, room and board are $4000/yr, and expenses are $3500/yr. A student loan for $2500/yr is available, which your aunt and uncle are willing to assume after your cousin graduates. Your cousin believes (but is not sure) that she can earn up to $4000/yr with an off-campus job and summer employment. Equivalent educations will be received at both universities.

   The decision seemed to be an easy one until your cousin found out that the state university will not accept all her units and that graduation will be delayed by one semester (quarter). Your cousin can potentially earn $25,000/yr after graduation but does not mind delaying her entry into the professional world. Write a feasibility report for your aunt and uncle to recommend a course of action.

2. The four executive officers of your student organization are expected to attend the annual planning conference at the beginning of each school year. The cost for each of the four attendees varies from $200 to $600, depending on the location. Even though your organization reimburses each attendee for 50 percent of this cost, several officers have been unable to attend this conference because the cost is prohibitively expensive.

   At one of your board meetings, it was suggested that your organization could raise funds for the remaining 50 percent by purchasing an electric grill and selling hot dog lunches in the quadrangle of your campus. The grill can be purchased for $800 from your treasury. A hot dog lunch presently sells for $2.50, which has a net profit of $1.00. It is estimated that 75 lunches can be sold each day that the grill is set up. The board was very enthusiastic about this proposal; therefore, you volunteered to investigate the feasibility of the proposal and to write a report.

After a few telephone calls, you find out that each student organization is limited to two fund-raisers per year on campus. However, you can rent your grill to any of the other 22 student organizations. It is estimated that the other organizations will pay up to $25 for each rental; however, unless you advertise in the school newspaper, it is uncertain that any other organization would be interested. An ad costs $5 and will run for 1 week. Because of liability insurance, school regulations require that you store the grill on campus and that it may not be removed except for repair. Write your feasibility report, with a recommendation, to present to the executive officers of your organization.

3. Your department at school has four tensile testers in the strength of materials laboratory, which, because of their age, cost an average of $1000/yr each to maintain and repair. A local engineering company has offered to replace these testers with new, state-of-the-art machines at no cost in exchange for free tuition for up to 10 part-time students for the next 6 years. Tuition at your school is $40/unit, and the average part-time student completes 12 units/yr. The administration was not happy about foregoing potential income for the next 6 years until it realized that these students also buy books, eat meals on campus, buy parking stickers, and so on.

Write a feasibility report, and make a recommendation for presentation to the administration of your school.

Multiple Proposed Courses of Action

4. Your neighbor's daughter has been accepted to the college of her first choice for the fall semester. Her parents will be paying all her tuition and living expenses for 4 years and would like to minimize the cost of her living arrangement. There are several alternatives:

- Live in the dorms. Room and board cost $5000/yr. She would not need a car.
- Rent a room off campus for $250/mo for 9 mo/yr. Food would cost $150/mo. Because no public transportation is available, a used car would be purchased for commuting to campus. A reliable car costs $100 down and has payments of $97/mo for 4 years. Insurance and maintenance are $125/mo. The added expense of the car can be offset by more flexibility in finding summer employment in a high-paying industry.
- Buy a four-bedroom house adjacent to campus for $250,000. Payments are $1750/mo with $50,000 down, taxes are $3000/yr, and maintenance and utilities are $300/mo. Each of the three extra bedrooms can be rented to a student for $300/mo for 9 mo/yr. The daughter would be the prop-

erty manager. It is expected that the house can be sold at the time of the daughter's graduation for $275,000. A car would not be needed. Assume that the $50,000 down payment can earn 10 percent interest, that is, $5000/yr, when invested elsewhere. Ignore the effects of income tax.

Write a feasibility report advising a course of action for your neighbor.

5. The planners of a government agency have proposed three plans for a recreational area at a scenic location:
   - Build 50 picnic sites, each with a table, barbecue, and parking space.
   - Build 40 picnic sites and 25 camping sites.
   - Build 25 picnic sites, 15 camping sites, and 8 cabins.

   Each picnic site costs $1200 to build, $150/yr to maintain, and will be used approximately 200 times per year. Each camping site costs $2400 to build, $200/yr to maintain, and will be rented approximately 100 times per year at $2/day. Each cabin costs $5000 to build, $400/yr to maintain, and will be rented approximately 20 weekends/yr at $15/weekend. The recreational area is planned for a 20-year period. The three plans are equally attractive to the administrators of the agency.

   As a planner for the agency, you are asked to write a feasibility report to recommend the least expensive course of action. (Ignore the time value of money and inflation.)

6. To increase profits, your company is proposing to open a field office for the next 5 years. Three sites are available that meet your company's needs. Each site has a different purchase price, will generate different annual revenues, and will have different resale value in 5 years. Site A, in a decaying area, costs $150,000, will generate $18,000/yr in revenues, and has an estimated resale value of $140,000. Site B, in the commercial district, costs $178,000, will generate $13,000/yr in revenues, and has an estimated resale value of $190,000. Site C, overlooking the river, costs $184,000, will generate $12,000/yr in revenues, and has an estimated resale value of $200,000.

   You are asked to compare the three sites and prepare a feasibility report to submit to management. You may include intangible factors such as demographics and prestige of location, but ignore the effects of income tax, the time value of money, and inflation.

CHAPTER 19

Environmental Impact
Statements and Reports

*The effects of a project on the surrounding environment are the
subjects of environmental impact statements and reports.*

To preserve the natural environment, in 1969 the federal government passed the
Natural Environmental Policy Act (NEPA), which requires any federal agency
whose proposed activity can disturb the natural environment to file an environmental impact (defined as a measurable environmental event) statement.
Legislation in most states requires any private developer whose proposed activity
can disturb the natural environment to file an environmental impact report. These
written statements and reports respond to these legal requirements.

The regulatory agencies (e.g., the planning commission) and bodies (e.g., the
school board) that approve the type of activities that require environmental
impact statements and environmental impact reports typically have guidelines of
topics to be addressed in these statements and reports. The items discussed in this
chapter are typical of these guidelines.

ENVIRONMENTAL IMPACT STATEMENTS AND ENVIRONMENTAL IMPACT REPORTS COMPARED WITH FEASIBILITY REPORTS

Environmental impact statements (commonly known as EISs) and environmental
impact reports (commonly known as EIRs) are very similar to feasibility reports
because, similar to feasibility reports, they evaluate potential courses of action. In
fact, many writers refer to them as feasibility reports. Hereafter, except for the section entitled, "Environmental Impact Statements Compared with Environmental
Impact Reports" beginning on page 282, EISs are included in all references to EIRs.
Because of the difference of the perspectives of EIRs and feasibility reports, EIRs are
treated separately in this text. Also, EIRs evaluate environmental (rather than eco-

nomic or technical) factors; therefore, many of the subjects in EIRs are unique to these reports.

Feasibility reports discuss the potential courses of action from the perspective of management and clients who are uncertain of the practicalities of these courses, whereas EIRs discuss potential courses of action whose outcomes have already been determined to be practical by management and clients but whose effect on the environment is uncertain.

Example

Suppose a developer is considering the construction of apartments only two blocks from a railroad. The feasibility report, written before a developer has decided to proceed with a project, would address (among other considerations) the effect of the noise from the railroad on the reduced ability of these apartments to command competitive rents from prospective tenants and, therefore, the potential of the developer to profit from the project.

The EIR would be written after this feasibility report demonstrated that these apartments were economically feasible for the developer, even at the reduced rents, and the developer has decided to proceed with the project. This EIR would probably conclude that the effect of the additional noise that is due to construction and increased traffic as a direct result of these proposed apartments is negligible in the already noisy environment. Although the prospective tenants may not approve of the existing noise from the railroad, the EIR does not address this topic except to the extent that reduced rents may attract lower-income tenants who affect the surrounding environment in other ways (e.g., they may drive older vehicles that have greater hydrocarbon emissions).

ENVIRONMENTAL IMPACT STATEMENTS COMPARED WITH ENVIRONMENTAL IMPACT REPORTS

EISs (written to comply with federal regulations) and EIRs (written to comply with state regulations) are identical in purpose, audience, and content, with one exception. Because EISs are for compliance with federal rather than state regulations, the scope of coverage is more global than for EIRs. For example, if a proposed project increased the surface runoff from rainwater, an EIR would address the path of this excess runoff into a river or channel and the effect of the excess runoff on this river or channel. An EIS, however, would also address the effects on the riparian (water) rights of the downstream neighbors, regardless of the jurisdiction for these rights, that are due to this excess runoff in the river or channel.

PREPARATION OF EIRS

Regulatory government agencies often prepare their own EIRs; sometimes however, the report is prepared by a consultant who is a specialist in environmental impact issues. The preparers of EIRs usually have backgrounds in engineering, planning, landscape architecture, and the physical sciences.

Clients are obviously biased and would like their projects approved. However, when you write EIRs for clients, these EIRs must be objective and without regard to your clients' viewpoints. When the government agency has any reason for denying your clients' requests for permits, your clients want to know before they invest more time and money in the project.

Because the readers of EIRs may include professionals with nontechnical backgrounds, these reports should be written in plain language and without jargon. Graphics should be included whenever possible to provide readers with a clear mental image of the project and to demonstrate concepts. The contents of these reports are often orally presented at hearing and meetings.

PURPOSE AND AUDIENCE

The appropriate government regulatory agencies and bodies review the anticipated environmental impacts reported in EIRs. These reports are advisory only. When the environmental impacts do not violate the master plan (for cities and counties) or the environmental policy of the jurisdiction, a permit for construction may be issued. Occasionally, however, because of public policy (a demonstrated benefit for the general public), a permit is issued in spite of violation of the master plan or environmental policy of the jurisdiction.

CONTENT

EIRs evaluate the short-term and long-term effects of the project on the environment. Short-term effects of a project on the environment include factors that only temporarily affect the environment (e.g., the construction process of the project); factors that will subsequently diminish or disappear as a result of changes in technology (e.g., increased fuel-efficiency of vehicles); demographics (a history of people leaving the rural area for more industrialized towns and cities); and further development (e.g., the planned construction of a reservoir).

Long-term effects of a project on the environment include factors that are permanent (e.g., loss of a natural resource) and factors that will subsequently worsen (e.g., accumulation of toxic waste).

Occasionally, one environmental factor will have a significantly greater impact on the environment than the others, or a government agency or body is responsible for the regulation of only one potential environmental impact of a proposed project. To study that impact, a report may be required to address only that one factor. This report is identical to an environmental impact report except that the scope is limited to a single factor of the environment. The title of this report includes only that one factor (e.g., traffic impact report, noise impact report, air quality impact report).

Note: Undoubtedly, you will use the words *affect* and *effect* in your reports: words that easily are interchanged incorrectly. *Affect,* usually a verb in the context of EIRs, is synonymous with *influence* and ordinarily precedes the event.

Effect is synonymous with *result* or *outcome* and ordinarily is the result of the event.

Example

Correct: Building the dam across the river will *affect* [i.e., influence] the users downstream.

Correct: However, the *effect* [i.e., result] of the dam on these users cannot be predicted without further study.

To express already determined changes, use directional words rather than *affect* and *effect*, which are vague unless further explanation is included.

Example

Vague: The air filtration system will *affect* the air quality.

Better: The air filtration system will *improve* the air quality.

Vague: The loss of wildlife will have an *effect* on the vegetation.

Better: The loss of wildlife will *increase* the vegetation.

Environmental impact statements and reports typically include the following sections (Figure 19–1):

1. *Abstract:* ordinarily called an executive summary in these reports. An informative abstract that summarizes the information in the report including the conclusions, recommendations, and alternatives.
2. *Introduction:* a statement of the proposed project. Includes the authorization of the report and identifies the regulatory agency or body that will review the report.
3. *Scope:* a statement of the factors that are studied. Identifies any factors that are not included in the study.
4. *Description of the proposed project:* details that help the reader understand the project. Includes area maps and photographs of the existing land area and layouts of the completed project.
5. *Analysis:* an analysis of the impact of the proposed project on the environment. Considers at least the following suggested items:

    Soil
    Geology
    Vegetation
    Biology
    Wildlife
    Water and watersheds
    Climate
    Air quality
    History and archaeology
    Transportation
    Demographics
    Noise
    Natural Resources

FIGURE 19–1

Sections of an EIR or EIS

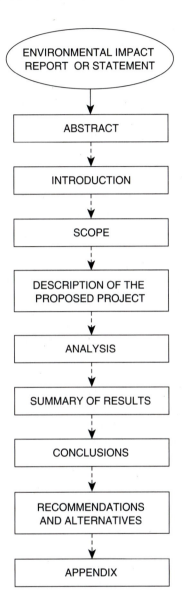

The analysis section introduces the data and information used as a basis for the analysis and to form the conclusions, recommendations, and alternatives. Information must be clearly explained. Includes calculations with the data when applicable.

Original data sheets and research information that are used in the analysis of this section are often included in the appendix of the report. (For your student assignment, you may need to develop these data and information.)

Because this section discusses many items and is the basis for the ending of the report, it is usually lengthy.

6. *Summary of results:* an abbreviated form of what is determined in the analysis section of the report. Uses tables and charts to demonstrate these results.

7. *Conclusions:* the natural outcome from the data and information presented in the analysis and summary of results. This section is the basis for the recommendations and alternatives.

8. *Recommendations and alternatives:* mitigation measures (recommendations) to minimize the impact of the proposed project on the environment, and alternative projects for further study that may achieve the same objective as the proposed project. Recommendations and alternatives are frequently included in the same section because the difference between a recommendation and an alternative may only be a matter of degree (i.e., a recommendation of sufficient size may be considered to be an alternative).

9. *Appendix:* data sheets and information supporting the analysis.

Sample Transportation Analysis

Valley Boulevard and Temple Avenue are both heavily traveled streets. Presently, these streets have an average daily traffic of 7500 vehicles and 6200 vehicles, respectively. This traffic is anticipated to increase by 6.2 percent on Valley Boulevard and 3.8 percent on Temple Avenue as a result of the shopping center.

Critique of Sample Transportation Analysis

This section appropriately includes the data for the anticipated increase in traffic. However, the basis or source of these data is unknown. Were the data based on calculations? If so, the calculations should be shown. When the data were based on research by a consultant, the research should be included either in this section or in the appendix.

Appropriately, no judgment is made concerning the effect of this increase in traffic. Very often, several environmental effects are integrated to form a cumulative impact. Judgments concerning the impact of the individual factors therefore should be reserved for the conclusion section of the report, where the writer can make a holistic judgment.

SAMPLE ASSIGNMENT—ENVIRONMENTAL IMPACT REPORT

See the instructions for the Student Assignment beginning on page 292. The facts of the sample assignment regarding a proposed project are as follows:

Athletic Stadium

A new 40,000-seat athletic stadium is proposed for athletic and social events on the campus of [*insert name of your school*] in [*insert name of the city*]. Donations from interested alumni and local industry will make the construction of this stadium and 20,000-car parking lot possible. It will be located on the [*northeast, northwest, etc.*]

corner of [*insert the intersection*] on 80 acres of vacant land that are owned by the city and are available on a 50-year lease. It is anticipated that rental expense of the land can be met from future donations from alumni and by renting the stadium to any interested groups when it is not being used by the school. The stadium will also be used by physical education classes and will provide flexibility for the school to expand its physical education program.

Several seating sections including bleachers will be available. Locker rooms for the athletes will be available under the stands. Appropriate facilities will be provided for press and media coverage. Refreshment stands will be located at convenient locations and will be maintained and run by student organizations as fund-raisers. Proceeds from events and parking will be used to maintain the facility by contract personnel. Profits will be used for academic programs.

Construction of the facility will take approximately 6 months and should not disturb any school activity.

Figure 19–2 is an illustration of this type of EIR.

Critique of Sample Environmental Impact Report

The sample in Figure 19–2 does not have a table of contents to help the readers locate the sections of interest. Short reports written for a homogeneous audience do not need a table of contents; however, EIRs are always of sufficient length and have a sufficiently diverse audience to require it. In particular, this student sample is of sufficient length to include a table of contents.

The sample is missing an abstract, which is the most widely read section of the report for nontechnical readers. The absence of this section reduces the effectiveness of this report.

Also, this sample is missing a scope. The section alerts readers to which factors are studied and to the relevance of the report to the reader's requirements.

The summary of results, as addressed in this report, would be more suitably entitled "Conclusions," because judgments concerning the impacts are included. A list with each item preceded by a bullet would simplify understanding the information.

A more appropriate summary of results for this report would include a quantitative review of the data and information presented in the analysis for easy comparison by readers. Tables are usually used for this purpose.

The recommendations and alternatives section of this report begins with an alternative site location of the stadium to preserve the scenic approach to the campus. The third paragraph is concerned with the use of the stadium to minimize the traffic impact. These recommendations appropriately consider the use as well as the design of the stadium.

The recommendations and alternatives are included near the front of the report for greater visibility. However, this section should be in a list for easy understanding.

The analysis appropriately includes data and references to the appendix. This section judges the effects of the data only when the impacts are negligible. Although judgments are usually reserved for the conclusions section of the report, this is satisfactory for judgments concerning items with negligible impacts. This section of the report should include calculations when applicable.

ENVIRONMENTAL IMPACT REPORT
FOR THE VIRGINIA STATE POLYTECHNIC STADIUM PROJECT

Introduction

EIRs always include an informative abstract to summarize the findings in the report.

The construction of an athletic stadium is proposed for Virginia State Polytechnic University at Roanoke (State University). This document addresses the anticipated impact that this project will have on the environment. It is prepared in accordance with the Virginia Environmental Quality Act. This document identifies potential impacts for this project and suggests possible mitigation measures.

Project Description

A scope section should precede the project description in an EIR.

The proposed athletic stadium will have 40,000 seats and parking space for 20,000 cars. The site includes 160 acres on the northwest corner of Glendale Boulevard and Church Avenue in the City of Roanoke (see Fig. 1) [omitted for space considerations]. The property is owned by the Virginia State Polytechnic University and is currently used by the Agriculture department of the university for the production of feed crops. The stadium will be 900 feet long, 450 feet wide, and 66 feet high (see Fig. 2) [omitted for space considerations] and will be used by State University's intercollegiate athletic and physical education programs for school events. When not used by the school, it will be available for rental by off-campus groups. There will be a 5000-space, two-story parking structure and a 15,000 space parking lot with access to Glendale Boulevard will accompany the stadium. Construction of the facility will require approximately 6 months.

FIGURE 19–2
Sample Environmental Impact Report (pp. 288–290)

The summary of results would be more suitably entitled "Conclusions" in this report, because judgments concerning impacts are included.

Summary of Results

The construction of an athletic stadium along the school's southern boundary would create traffic congestion on fronting streets. Because stadium events typically occur on weekends, when the volume of traffic on these streets is at a minimum, only limited impact on traffic flow in the area should result. Weekend use will also minimize the disturbance created by increased noise on the surrounding community.

Studies of the impact on soil, vegetation, climate, geology, water, and wildlife indicate no adverse effects from the stadium.

Recommendations and Alternatives

The alternative of no stadium on the proposed site would preserve the scenic approach to the campus but not fulfill the needs of the school.

An alternative site might be an area directly west of the Swine Research unit of the school's Livestock department. This site is comparable in size to the Glendale Boulevard location. The natural contours of the area will minimize excavation.

Due to traffic congestion on Church Avenue during commuting hours, events that create additional traffic on this street should be scheduled to avoid commuting hours.

Analysis

Transportation

Glendale Boulevard is a four-lane primary road between Roanoke and the city of Radford. Church Avenue is a four-lane secondary road passing through the cities of East Walnut, Roanoke, and Phillip's Ranch. The traffic flow on each of these roads is greatest during the morning and afternoon commuting hours.

The traffic study commissioned for this report (see Appendix A) determined that the level of service on Glendale Boulevard is constant at Level A throughout the day, every day of the year. The level of service on Church Avenue is "A" during non-commuting hours but "F" (below minimum standards) during the commuting hours of 7 to 9 a.m. and 4 to 6 p.m.

Approximately 20 trains a day use a rail system parallel to Glendale Boulevard. The waiting time for vehicles averages approximately 5 minutes and would not significantly affect traffic flow.

Vegetation

The proposed stadium site is currently used by the university's Agriculture department for the production of feed crops and to grow pumpkins for various school districts in the area. Sycamore trees are at 50-foot intervals along Church Avenue in the right-of-way. No unique or endangered plant species are known to be growing on the proposed site.

Climate

Because of the elliptical shape of the proposed stadium, the wind patterns in the area would not be significantly affected. On calm days, ground temperatures can increase an average of 5 degrees Fahrenheit during the day, which is due to heat absorption and convection of the asphalt parking lots. The local humidity would be reduced an average of 12 percent. Because the site is in a foothill valley connecting the Roanoke and Blue Ridge basins, it experiences breezes of 5 knots or more 318 days per year. The local climate would not be affected on those days.

[For space considerations, the remaining sections of the analysis have been omitted. These sections were entitled "Soil," "Geology," "Noise," "History and Archeology," "People and Psychology," "Air Quality," "Wildlife and Biology," and "Water and Watershed."]

[Figure 1, a map of the campus and surroundings, and Figure 2, a computer-generated drawing of the proposed stadium—omitted for space considerations—would be placed here.]

[For space considerations, the Appendix is omitted]

FIGURE 19–2
(*continued*)

KEY CONCEPTS

- Readers of EISs and EIRs include officials with nontechnical backgrounds. Therefore, it is essential to avoid jargon and to write your report in plain language.

- Many readers use the abstract (or executive summary) exclusively to gather the knowledge for their opinions. Therefore, the abstract must summarize the conclusions, recommendations, and alternatives clearly.

- The information in EISs and EIRs is frequently presented at hearings and meetings. Sometimes this information is used to persuade a public audience. Therefore, graphics in your reports enhance their utility.

STUDENT ASSIGNMENT

You are employed by an environmental consulting firm. Your client is requesting a permit for a project that affects the natural environment and employs you to write an EIR.

Your assignment is to respond to the proposed project shown in Figure 19–3. Because projects below a minimum size are exempt from the EIS and EIR requirements, these student assignments would probably be exempt in most jurisdictions; however, these assignments exemplify the types of projects that require EISs and EIRs to be written.

You have received the letter in Figure 19–3 from your client confirming your contract to write an environmental impact report.

Proposed Projects

1. **Duck pond**   A duck pond is proposed for the enjoyment of faculty, staff, and students on the campus in [name of the city]. It will be located at the [insert northeast, northwest, etc.] corner of [insert the intersection] and will be in the shape of an oval approximately one hundred feet long by seventy-five feet wide. The maximum depth will be 3 feet at the center. The area surrounding the pond will be landscaped with trees and bushes. A 5-foot-wide walkway will lead from the street to eight benches and two picnic tables surrounding the pond and located approximately 15 feet from its edge. A 3-foot-wide walkway will encircle the pond at approximately 10 feet from the edge of the water. Construction of the pond will be during the summer break, and completion should require approximately 2 months.

   The pond initially will be stocked with 20 ducks and approximately 100 fish. The pond and surrounding area will be maintained by the school landscape maintenance department. The wildlife will be cared for by a state-licensed wildlife preservationist on a semiweekly basis.

YOUR UNIVERSITY
123 CAMPUS DRIVE
SMART CITY, AT 56789
(800) 444–5555

January 2, 1996

[your name]
Environmental Services, Inc.
123 Clean Place
Aseasy, AS 12345

Dear Mr./Mrs. [your name]

[Your School] is presently in the process of obtaining the permit for [name of the project and its approximate location on campus]. We will need an investigative report of the potential environmental impacts before this permit is issued. A brief description of the project is enclosed.

Submit the report to me on [due date] for presentation to the City of [name of the city] Planning Department. You may submit your bill at $50 per hour plus expenses with the report.

Please call me if you have any questions.

Very truly yours,

Hy A. Chiever

Hy A. Chiever
Professor
School of Engineering

Encl.
HC:jnb

FIGURE 19–3
Project Proposal for a Student
Assignment

2. **Student housing** A student housing development is proposed at [your school] for married students on the [insert northeast, northwest, etc.] corner of [insert intersection]. The development will consist of efficiency and one-bedroom apartments on the land presently used for equipment storage by the maintenance department of the school.

The three 3-story buildings will face to the west and overlook an Olympic-size pool, a kiddy pool, a playground, and landscaped walkways and lawns. The building line for the back of the three buildings will be 25 feet from the westerly edge of the 250-car parking lot adjacent to [insert name of the street]. The parking lot will be 100 feet wide for the full length of the development and have an area reserved for recreational vehicles. The [north, south, east or west] side of the most [northerly, southerly, etc.] building will be 150 feet from [insert name of the street]. The buildings will be spaced 55 feet apart and will be in a row.

Each 50-foot by 150-foot building will have 16 units per floor: nine efficiency units and seven one-bedroom units. Each building will have a storage room and gym on the first floor, a lounge on the second floor, and a study on the third floor. In addition to a front entrance, each building will have a rear entrance from the parking lot.

The rent is planned to be $250 per month for the efficiency apartments and $300 per month for the one-bedroom apartments. Efficiency units will not be rented to students with children. Students with children more than 1 year of age will live on the first floor. Each building will have an elevator easily accessible from the parking lot.

3. **Student tram** Because of the expansion of the [your school] campus in [name of the city] and the increased distances from the student parking lot to the class buildings, a two-car, open-air tram has been proposed to transport students around campus. The tram will stop at designated points in the parking lot and at all major buildings and facilities. Its route is planned to transit the perimeter of campus as much as possible to avoid the populated areas. The 2-mile, 15-minute route will originate at the far end of the parking lot every 20 minutes beginning at 6:40 A.M. and will continue until 11:00 P.M.

Each car will have a continuous bench seat that faces out on each side of the tram for easy entry and exit by students. There will be no guard rail. It will have a weather cover for protection from the sun and rain; the tram will not operate during snowstorms. Each car will have a capacity of 40 students. The tram will be powered by a 40-horsepower gasoline engine and will have a maximum speed of 35 miles/hr. The drivers will be school employees and licensed by the state to drive school busses.

4. **Staff parking lot** A staff parking lot is proposed for the use of administrative personnel employed in the new university office building on the [name of your school] campus in [name of the city]. The 100-foot-wide by 150-foot-

long parking lot will have approximately 50 parking stalls (including four handicapped stalls).

The lot will be bordered on the [north, south, east, or west] by [insert the name of the street] and on the [north, south, east, or west] by the office building. Access to the lot will be by key-card from [insert the name of the street]. There will be a pedestrian entry from the parking lot to the rear entrance of the office building. The lot will be esthetically landscaped with shrubbery and trees and separated from the street by a 12-foot-wide grass parkway. The slope for drainage of the asphalt lot will be 4 percent.

5. **Snack shop**   A 2000-square-foot snack shop is proposed on the [name of your school] in [name of the city] for the enjoyment of students, faculty, and staff. Pastries, desserts, sodas, and sandwiches will be sold at prices that are competitive with the school cafeteria. Tables and benches will seat approximately 75 students inside the shop and another 50 students near a take-out window. The facility will be self-service. Pastries will be baked on the premises and will be fresh every day. The shop will be managed by contract personnel; however, all employees will be full-time students in need of financial aid. Profits will be used to subsidize minority and academic enrichment programs on campus.

The shop, located [insert location] near the center of campus, will have a wood-frame structure with a tile roof. The interior will be attractively decorated, and the exterior will be attractively landscaped.

Deliveries will be made and trash picked up before classes begin in the morning. In the event that student noise disrupts nearby classes, a sound barrier will be erected.

CHAPTER 20

Technical and Sociotechnical Articles

Professionals and laypeople keep current by reading technical and sociotechnical articles.

Professionals frequently have the opportunity to disseminate their knowledge to others by writing articles. Typically, professionals write these articles for one primary reason: self-satisfaction. However, the author of these articles will also experience professional growth, peer recognition, increased credibility, and perhaps, job security. This chapter discusses these articles.

TECHNICAL ARTICLES

Technical articles disseminate information to professionals practicing in the same, or related, fields of expertise as the writers. These articles can be found as published articles in technical journals, magazines, and newsletters of professional organizations, and as articles written for the edification of colleagues and associates within a company or agency. They are usually written by professionals with a technical background.

SOCIOTECHNICAL ARTICLES

Sociotechnical articles disseminate information having a technical basis but whose primary concern is the effect of technology on our quality of life. Sometimes these articles are intended to persuade readers to a preferred point of view. These articles may be read by other professionals and the general public and are usually published in technical journals, magazines, and newsletters of professional organizations, as well as newspapers and periodicals.

Sociotechnical articles and editorials may be written by professionals with backgrounds in science and technology, social science, political science, and occasionally business.

"White papers" state the positions of professional societies such as the National Society of Professional Engineers concerning sociotechnical issues, for example, safe disposal of toxic waste. These papers are written by select technical committees appointed by these professional societies and presented to our lawmakers in the state capitols and Washington, D.C., to influence legislation.

TECHNICAL AND SOCIOTECHNICAL ARTICLES COMPARED WITH REPORTS

Reports communicate information to targeted audiences that need to comprehend the included information. On the other hand, even though technical and sociotechnical *articles* are intended for targeted audiences, these audiences have no need to comprehend, or even read, the information included in these articles.

Often, articles have the same focus as reports: activity and product evaluations, technical and economic feasibility studies, research, and research and development. Published articles with these focuses differ from reports only in the following ways:

- Writers must attract the attention and maintain the interest of the potential readers.

- Because the scope of the audience is frequently broader for articles than reports, in these articles, technical terms and jargon may need to be defined.

Other than these differences, the content of these articles is the same as for reports. To publish articles that have the same focus as reports, please see the appropriate chapter in this text for the content, and then read this chapter for a discussion of the preceding bulleted items.

Articles that have the same focus as reports and that are intended to be read by colleagues and associates in your organization may not necessarily differ from reports, because the readers are concerned with the information included in these articles.

PERSUASIVE ARTICLES COMPARED WITH DESCRIPTIVE ARTICLES

Descriptive articles discuss their subjects with objective points of view. Much of the text is given to descriptive narrative and *objective* evaluation of processes and events.

To be effective, persuasive articles must also describe the processes, events, and situations that are the subjects of these articles before making any *subjective* evaluations. However, before readers will accept these evaluations made by writers, these readers must be convinced the writer is using a rational approach to the evaluation. Therefore, to convince readers, substantiate the preferred points of view in persuasive articles with facts and data rather than with inflammatory statements intended to arouse the emotions of readers.

Example

Inflammatory: The increase in noise at this location from the additional traffic will undoubtedly affect the ability of students to concentrate and learn.

Subjective: Children who attend schools located in close proximity to freeways consistently score between 6 and 10 percent lower on achievement tests than other children of similar backgrounds who attend schools that are remote from such freeways.

The conclusion should be a natural result of these facts and data that are presented.

The primary subjects of this chapter are descriptive and persuasive articles concerning newsworthy events, rather than articles that have the same focuses as reports.

PURPOSE AND AUDIENCE

Writers of technical and sociotechnical articles usually have experienced, or have knowledge of

- A newsworthy technical event (e.g., completion of a major construction project, or the first flight of an experimental aircraft).

- A newsworthy current event concerning a technical or sociotechnical issue that has public interest because of its emotional appeal (e.g., an earthquake, or change of zoning laws).

- A technical invention, discovery, or update (e.g., a new type of motor oil for extending the life of automobile engines, a new element found in the earth, or a status report on semiconductor technology). These articles may have the same focuses as reports or also may treat the subject as newsworthy events.

- A solution to a short or long-range sociotechnical problem (e.g., the effectiveness of restricting use of freeways by trucks during rush hour, or disposal of nuclear waste). These articles may have the same focuses as reports, or may be written as descriptive or persuasive articles.

Colleagues, associates, legislators, and the general public may be interested in reading these articles and opinions to keep up with the state of the art, and to keep informed. The readers of these articles include engineers, scientists, social and political scientists, businesspeople, and the general public.

CONTENT

General

Because the purpose of an article is to disseminate information, before beginning to write an article, the author must ask, "What information do I want the reader to know?" When this has been determined, the content of the article can be presented in a manner that assists the reader to understand this information.

Headings are usually used for technical articles. However, when an article is essentially a chronology of events or a description of a project, and the article is short, many writers prefer not to include headings, which may break the concentration of readers and cause them to lose interest.

However, although articles concerning newsworthy events and persuasive articles may include headings, they usually do not include tables of contents, because the writers intend these articles to be read in their entirety.

To increase the readability of articles, use graphics with captions to help the readers visualize the concepts discussed. These graphics are placed following the references to them in the text and preferably on the same page.

Published Articles

Technical and sociotechnical articles in publications compete for the attention of readers. Therefore, to have the desired impact, these articles must attract the attention and maintain the interest of the intended readers.

Titles to these articles need to be descriptive and have an emphasis to give the reader a reason to read the article. For example, rather than titling an article "Increased Strength of J6 Steel"—a nondescriptive title that lacks interest—title it "Heat-Treating Increases the Strength of J6 Steel by 18%." This describes the event by emphasizing the effect of a procedure. Typically, titles are active statements rather than passive sentences (i.e., not "The Strength of J6 Steel Is Increased by 18% by Heat-Treating"). Notice the % symbol—discussed elsewhere in this text to be used primarily in calculations, figures, and tables—is used for emphasis in these titles and to capture the attention of the audience.

Titles of persuasive articles are often phrased as questions that may immediately tell the potential readers the biases of the writers and, therefore, dissuade readers without these biases from reading these articles. For example, an editorial concerning the increased strength of J6 Steel might be entitled "Is the Increased Strength of J6 Steel Really Cost-Effective?" This title, by inclusion of the word *really*, places doubt on the practical value of this increased strength. Also, by not including the amount of the increased strength, it subtly negates its importance.

To keep the attention of your readers, the opening sentence needs to promise information not available elsewhere.

Example

Poor opening: The strength of J6 steel is increased 18% by heat-treating.

Better: Contrary to the belief that the strength of J6 steel cannot be increased by heat-treating as a result of its low carbon content, recent tests have demonstrated heat-treatment to be effective in a nitrogen-rich environment.

The opening phrase "Contrary to the belief. . ." piques the curiosity of readers. The remainder of the sentence tells the readers something that they are probably not aware of and would hopefully like to know.

Publications usually have specific guidelines for articles to be considered for publication. These guidelines can be obtained by contacting the editor of the publication, but they may also be explained in the publication.

Intraorganizational Articles

In contrast with published articles, which compete for the attention of readers, your colleagues and associates in a company or agency have pre-established interests in the subjects discussed in the articles written exclusively for their benefit. Their interest in reading these articles can be created with cover memos and letters explaining the relevance of these articles to current projects or by inclusion in company and agency newsletters, which are intended to keep employees current.

The following suggested headings (or types of information to be included when headings are not used) are general because the intent and content of articles can vary considerably. However, use the following as a guideline (Figure 20–1):

1. *Title:* a brief description of the contents of the article. It is intended to catch the attention of potential readers by emphasizing something concerning which they would probably like more information.

2. *Byline:* your name.

3. *Abstract:* a descriptive abstract (see Chapter 7) telling the potential readers what is discussed is essential for any published article. Intraorganizational articles usually include an abstract in the cover letter or memo.

4. *Definition of terms:* depending on the intended audience, terms and jargon may need to be defined. Because definitions of jargon are not easily obtainable elsewhere, any that are used must be defined.

5. *Introduction and background:* the history and purpose of the process, event, or situation.

6. *Description of the subject*: description and evaluation of the process, event, or situation. For persuasive articles, this section must include sufficient details for the readers to come to the desired conclusion.

FIGURE 20–1
Components of a Technical
Article

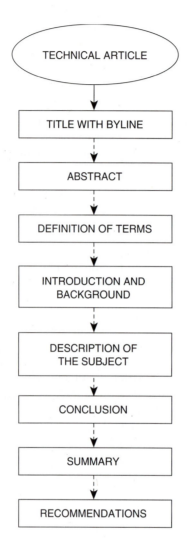

7. *Conclusion:* what can be concluded from the description of the subject. This section states the findings based on the facts and data presented. The conclusion section is the essence of persuasive articles.

8. *Summary:* a review of the details and a brief forecast of the anticipated effects resulting from the process, event, or situation.

9. *Recommendations:* possible applications; possible solutions included in articles concerning sociotechnical issues. These recommendations must be a natural consequence of the facts and data presented in the article and may reflect the viewpoint of the writer.

Samples of professionally written technical articles may be found in any scientific or engineering magazine.

Sample Title and Opening Paragraph of a Persuasive Article

Do We Need Another Freeway?

by Carolyn Kelsey

In an era when the voters are approving slow-growth and no-growth initiatives, Congress voted for the construction of a new freeway to connect the residential districts of our city with existing Interstate 20. The measure was barely approved by only two votes just minutes before Congress adjourned for the Christmas holiday.

Critique of Title and Opening Paragraph of a Persuasive Article

This article is clearly intended to evoke the emotions of readers. By using the word *another* in the title, it implies that another freeway is not needed. The opening sentence indicates that Congress is acting contrary to the wishes of voters. Mention of the marginal vote in the same sentence as the adjournment for the Christmas holiday implies that the vote was not seriously considered.

An article with this degree of persuasion in the title and opening paragraph would appeal to the emotions of readers and might be found in the editorial section of a newspaper or periodical.

However, the title would probably attract only those readers who were already emotionally involved with the issues of a new freeway. Because of its obvious bias, the title may not attract readers who are uninformed concerning this issue and who, if persuaded, may be potential opponents to this freeway.

To attract greater attention, this article could have been entitled "Freeway Approved by Congress to Connect With Interstate 20." This title includes details and arouses the interest of readers who want to know more.

To increase credibility, the introductory paragraph could have included facts and data concerning the approval of the construction and should not have mentioned the adjournment until later in the article, if at all. For example, the opening paragraph might have been,

> By a vote of 83 to 81, Congress approved the construction of a freeway to connect the residential districts of Joytown with Interstate 20. It is planned that bids for the design and construction of this proposed freeway will begin within 12 months.

Also, the paragraphs immediately following the introductory paragraph in this article could discuss statistics concerning present commuting times of the residents, population growth of recent years, and business development. Later paragraphs then could discuss the results of studies concerning the detrimental effects of the proposed freeway in Joytown.

Hopefully, with facts and data, the curious reader has been persuaded to the writer's point of view. The final paragraph might include a notice of a city council meeting to discuss this issue with the public or another activity that will occur concerning the construction of the proposed freeway.

SAMPLE ASSIGNMENT—TECHNICAL ARTICLE

See Student Assignment 3 on page 312, and use the following fact and data sheet:

Hydroelectric Plant

Bath County, Virginia

Virginia Electric Power Company

Completed in 1994

World's largest
  6 turbine/pump generator/motor units
  2100 mw

Two dams—rockfill/earthfill dam of upper reservoir
  Highest dam east of the Mississippi
  $1.6 million

Design began in 1982

License granted in 1986

Decrease in peak load growth rate after 1988

High cost of borrowing

Slowdowns, changes

Allegheny Power Systems became partner in 1989
  Preliminary exploration
  Satisfactory permeability

Adequate rock-bearing capacity

Seepage carefully monitored—400 electric piezometers

Rock faults, slope instability

Dwyideg anchors 150 ft long required in tunnel excavation

Modifications from original design—geologic conditions
  Upper dam site moved downstream
  Powerhouse at surface rather than underground
  Realignment of power water conduits
  Vertical rather than sloping conduits

Dams—impervious cores
  Upper dam—460 ft high, 2200 ft long, 17 million cu yd
  Lower dam—150 ft high
  38,000 cfs spillway—2 radial gates
  1:60 model tested at Georgia Tech

Powerhouse
  500 ft × 170 ft × 200 ft—reinforced concrete
  Walls must resist submergence in fluctuating reservoir

Main generator hall
        Six 65-ft bays—reversible turbine/pump generator motor
    Power conduits
        Three 28 1/2-ft-diameter tunnels between reservoirs
            18-in. concrete lining
        Two 18-ft-diameter penstocks to powerhouse

Figures 20–2 and 20–3 illustrate two different technical articles.

Critique of Technical Article—Sample 1

Sample 1 in Figure 20–2, appropriately includes a title that creates interest for general readers of a professional magazine by starting with "A New Era. . . ."

This sample includes figures with captions that, inappropriately, are not cited in the body of the text.

Headings are not included in this sample. For general interest articles intended to be read in their entirety, this is acceptable.

An abstract is not included. Although an abstract is not necessarily included with the published article for magazines intended for general readership, it is an essential element for an article to be considered for publication. Technical journals always include an abstract with each article.

This sample does not include definitions of terms, because it is expected that interested readers have a knowledge of hydroelectric power and would understand the meanings of the low-technology words.

Although some descriptive narrative with details is included in this sample, more details should be included for this sample to be a feature article. This sample appropriately discusses the background, economics, and technology of the hydroelectric plant.

The closing paragraph could be improved by including a summary of the article, or, more interestingly, a discussion of the potential future advantages that could result from the construction of the plant.

To be considered for publication, the author should include an abstract.

A NEW ERA IN HYDROELECTRIC POWER

By Robert Bigham

Completed in 1994 by Virginia Electric Power Company, the Bath County Dam is the world's largest hydroelectric generation plant in operation. This plant was conceived by the U.S. Army Corps of Engineers in 1982 as an alternative to the use of fossil fuels for electric power generation. Following a dispute with conservationists, the project was granted a license in 1986.

Although the estimated cost for the original design was $1.28 billion, the actual cost was $1.6 billion. During this time the project was revised to comply with the current seismic safety design standards as well as incorporate significant advances in turbine and electric generator design.

The Bath County project included the building of two dams. The upper reservoir, restrained by a rockfill/earthfill dam with an impervious core, required 17 million cubic yards of fill. The lower reservoir was restrained by a 150-ft high dam constructed of roller-compacted concrete.

Note that the figures are inappropriately not cited in the text.

Figure 1: The Bath County Dam Site During Construction
[Actual photographs and figures may not be required in student reports at the discretion of the instructor.]

The system boasts the world's largest generation system: 6 turbine/pump/generator motor sets capable of generating 2100 megawatts at peak load. The system is enclosed in a powerhouse constructed of reinforced concrete. The large size of the powerhouse posed numerous problems for structural engineers. The 500-ft long by 170-ft wide by 200-ft high structure required special analytical techniques to determine the strength required to overcome the substantial hydraulic forces imposed by submersion of the

1

FIGURE 20–2
Technical Article—Sample 1 (pp. 304–306)

powerhouse walls during fluctuation of the water level. Consideration of these forces convinced designers to redesign an early concept of an underground powerhouse. The final design of the main generator hall consisted of six 65-ft bays - one for each turbine set.

Water is fed to the powerhouse by two penstocks 18 ft in diameter. When rapid removal of the water in the upper reservoir is required, this water can be removed through three 28 1/2-ft diameter tunnels which connect the upper and lower reservoirs. Water can also be removed from the lower reservoir through a spillway at a flow rate of 38,000 cubic ft per second. Spillway flow is controlled by two computer-operated radial gates at the powerhouse.

A 1:60 scale structural and mechanical model for the Bath County Generation Plant and attached waterways was tested in George Tech before commencement of construction. This testing precipitated several major design changes as well as relocation of the upper dam site. Preliminary soils investigations also revealed satisfactory permeability and adequate rock-bearing capacity to allow a shift to the rockfill design. Earlier discoveries of several rock faults and slope instabilities at the original site made this relocation inevitable. These changes made it possible to realign the power conduits and construct them vertically rather than at a 52 degree slope used in the original design.

Figure 2: Revised Site Plan

Although project engineers are confident concerning the safety of the site and construction of the dams, the geological site conditions are continuously monitored to provide warning in the event of a failure. Four hundred piezometers are read by remote sensors and measurements are recorded

2

hourly by a central computer system. Continuous digital readout is possible at any terminal of the controlling computer.

The reversible turbine design is one of the attractive design features of the project. It has the capability to reverse three of the turbines during off-peak hours to pump water into the upper reservoir and raise the water level and thereby improve performance of the system during peak hours.

Even though this project boasts many superlatives, it was beset with problems from the beginning. Bid drawings were released near the end of the energy crisis. A decrease in peak loads that was due to energy conservation, combined with the high energy cost of borrowing, quickly diminished the urgency to build the project. This created a slowdown in the construction as portions of the system required modifications to reduce construction and operating costs. The project was even suspended for a short time while Department of Energy officials disputed the need for the plant. However, as fuel prices began to stabilize and energy consumption increased, the Department of Energy realized that this system was essential to the transition from a fossil-fueled economy.

Figure 3: Fossil Fuel Costs, 1978–1994

When construction was approximately 80 percent complete, Allegheny Power Systems purchased a 20 percent share in the project. In doing so, they became the first privately owned company to form a partnership with a department of the U.S. government.

3

FIGURE 20–2
(*continued*)

Critique of Technical Article—Sample 2

The title of Sample 2, shown in Figure 20–3, is mundane and unlikely to create much interest for potential readers. However, the abstract creates interest by discussing the subject of the article and telling potential readers the outstanding features of this hydroelectric plant. Appropriately, this abstract does not include details.

Sample 2 has two figures with captions: one is an aerial view, and the other is a cross section. The aerial view has no figure number and is not cited in the body of the text; sometimes this is done by experienced writers to create interest in articles that have either no other or many other cited figures. In this case, it is unusual that there are two figures, only one of which is numbered and cited.

Unlike Sample 1 (Figure 20–2), Sample 2 includes headings, which is more common for articles of this type. In addition, this sample includes interesting descriptive narrative with significant details.

The closing paragraph predicts the events that may be a direct consequence of the construction of this plant. The quote makes this prediction more memorable.

The byline is missing.

SUPER HYDROELECTRIC PLANT IN VIRGINIA

ABSTRACT

In August, 1990, construction began for the Bath County hydroelectric plant, which, when completed, will be the largest in the world. It will have six 65-ft bays that each house 2100-megawatt reversible motor units and two dams. One dam, constructed of earthfill, is expected to be the highest ever built east of the Mississippi River. The powerhouse will be built on ground level rather than underground, and will have walls made of heavily reinforced concrete. This project is expected to bring considerable attention to Bath County, Virginia.

Introduction

The world's largest hydroelectric plant, owned and operated by the Virginia Electric Power Company and Allegheny Power Systems, is located in Bath County, Virginia. It has the capability of producing over 2100 megawatts, but numerous difficulties have prevented it from fulfilling its potential.

The aerial photo should have a figure number and should be cited in the text.

Aerial Photo of Hydroelectric Plant in Bath County, Virginia
[Actual photographs and figures may not be required in student reports, at the discretion of the instructor.]

Description

This hydroelectric facility has two dams, two reservoirs, and six turbines, which are located in the lower dam. The upper dam, the larger of the two, is made of soil and rocks and has an impermeable core. It is 460 ft high, 2200 ft long, and contains more than 17 million cubic yards of soil and rocks. This dam retains the upper, larger reservoir, which releases water to the

1

FIGURE 20–3
Technical Article—Sample 2 (pp. 308–311)

lower reservoir through three 28.5-ft-diameter tunnels. Each tunnel required 150- ft long Dwyideg anchors to prevent the sides from caving in during the excavation, and to place the 18-inch-thick concrete lining.

The lower reservoir is retained by a reinforced concrete dam which is 500 ft long, 170 ft thick, and 200 ft high. It is divided into six 65-ft bays, each of which houses a turbine. Each of the six turbines contains a reversible turbine/pump connected to a generator/motor. Built into this dam is a spillway capable of releasing 38,000 cubic feet of water per second. Figure 2 shows a cross section through the dams.

Figure 2. Cross-Section Through Dams

Development

The Virginia Electric Power Company began design for this project in 1982. The preliminary exploration studies of the site indicated a satisfactory level of permeability and adequate rock-bearing capacity. There were, however, two main obstacles to overcome during its development.

The first obstacle was to obtain the necessary funds. The project engineers had estimated the construction cost to be $1.2 billion. After numerous meetings with various state and federal energy committees, as well as several banking and loan institutions, the required loans were secured.

The second obstacle was to obtain the necessary operating licenses and building permits. This required comprehensive environmental impact reports and many meetings with various city council members, state officials, and building departments. The license was finally granted in October 1986.

Construction

The construction began in January 1990, and had a targeted completion date for summer, 1992. However, there were several major problems. The

2

most minor of these was simply that the excavation process took considerably longer than the engineers had expected. But more importantly, the excavation at the site of the upper dam exposed considerably more faulting of the rock base than the preliminary site studies had determined. The faulting was so extensive that the upper dam had to be relocated downstream.

This relocation was the cause of many other revisions, including the tunnel locations and sizes, the reservoir cross sections, and the dam structure. All fractures in the rock base at the new site were grouted, and 400 electronically monitored piezometers were installed to check for seepage.

As a result of the many revisions and other delays, it became apparent that the target completion date would not be met, and the original $1.2 billion budget would be inadequate. As of April 1988, the construction was 80 percent complete, and the money for construction was nearly depleted. So in May 1989, Allegheny Power Systems became a 20 percent partner with The Virginia Electric Power Company and provided the necessary funds to complete the project. In May 1994, two years later than had been planned, the project was completed, at a total cost of $1.6 billion.

1994 to Present

As with any plant of this size, a few minor problems occurred with plant start-up. It took until early September 1994 for the reservoirs to become adequately filled to operate the turbines. As the turbine start-up procedure began, it became apparent that the inlets to the turbines were inadequately sized. This was a serious miscalculation by the engineers which required extensive redesign and retrofitting. This also strained the relationship between Allegheny Power and Virginia Power. Allegheny Power did not believe that they should provide the money to modify the design that was the result of inadequate engineering by Virginia Power. However, after much deliberation, Allegheny Power provided the additional funds, and in January 1995, the system became operational.

Unfortunately, since that time, the rainfall in Bath County and in most of the country as well, has been barely over one-half of the average. As a result, the turbines have not been able to be used to the fullest extent, and the power output of the plant has been only approximately 65 percent of the anticipated 2100 megawatts. This means less income and less profit for the power companies.

3

FIGURE 20–3
(*continued*)

The drought has also caused another problem: the company's relationship with the local communities. Generally, the local communities were not particularly happy with the plan to build a new plant. Although some people acknowledged the need, and recognized that more jobs would be available as a result, most people preferred that it not be built. And now that the river level is down, and water is becoming more scarce, they are blaming the new plant. Even though the new plant is not to be blamed for the drought, many people are using the plant as a scapegoat. It has created a feeling of ill-will between the power companies and the local people.

Another problem that recently became apparent is that the piezometers are indicating that the amount of seepage around and below the upper dam is increasing. When the dam was constructed, all the faults that could be located in the area were sealed with a watertight grout. This was a tedious, but necessary, job. The engineers now theorize that either some faults were not located, or the grouted joints have begun deteriorating.

The other possible explanation is that many areas in Virginia have veins of limestone in the substructure. Limestone is primarily composed of calcite, $CaCO3$, which is slightly soluble. Nearly all the caves in the area were once limestone veins that have since been dissolved by underground rivers. The engineers suspect the presence of a limestone vein that allows water to seep into the dam.

As seepage enters the interior of the dam structure, the dam erodes from the inside. If the erosion becomes sufficient to significantly weaken the dam, a failure could occur, which would be catastrophic. However, the seepage is carefully monitored, and the engineers assure us that preventative measures would be taken before failure is a possibility.

Conclusion

Although the world's largest hydroelectric plant has yet to produce to its capacity, it is being called a success by nearly everyone. Orel Munsen, spokesperson for the power plant, said, "All of the problems are behind us now. We have a great operation here and all we need is a couple of good, wet years to get everything back to normal, and then we can show everyone what we can do."

4

KEY CONCEPTS

- Technical articles are an opportunity for professionals to receive recognition by other professionals for the work they have performed.
- Technical articles are a means by which professionals employed in different organizations communicate with each other.
- Intraorganizational articles promote communication between departments.
- Articles by knowledgeable professionals intended to educate and persuade the general public provide long-range benefits for society otherwise not available.

STUDENT ASSIGNMENT

1. Write an article for a local periodical concerning a sociotechnical issue of importance in your community (e.g., traffic, smog, toxic waste). Include data from appropriate sources such as government and regulatory agencies. Although you may include recommendations as a result of the information presented, this article is not intended to persuade its readers.

2. Using either personal knowledge from part-time or summer employment, or an interview with a knowledgeable professional, write a technical or sociotechnical article for a technical periodical for one of the following:

   a. The construction of a private or public works project: Emphasize the technical challenges and accomplishments (failures) in the construction of this project. Consider design, size, purpose, cost, schedule, materials, inspection, compliance with government regulations and codes, politics, ownership, and employee relations.

   b. The manufacture of a product for use by industry or sale to the general public: Emphasize the technical challenges and accomplishments (failures) in the manufacture of this article. Consider design, schedule, costs, materials, production rates, quality control, workmanship and rejection rates, and compliance with specifications and industry standards.

   c. The design of a project or product by an engineering company or government agency: Emphasize the technical and administrative challenges and accomplishments (failures) in the design of this project or product. Consider design costs, analysis, technical expertise required, coordination with suppliers, concern for construction or manufacturing costs, concern for ease of construction or manufacturing, coordination of design groups, contractual compliance, and compliance with regulations, specifications, codes, and standards.

3. Assume you are writing an article for a scientific or engineering magazine such as *Scientific American, Engineering News Record,* or *Machine*

Design and have interviewed a researcher, designer, or the project engineer at a laboratory, an engineering office, a construction site, a manufacturing facility, or the site of a newsworthy event.

Use one of the following fact and data sheets (from a through f) to write an informative and interesting feature article for publication. Reorganize the facts and data on this sheet into a meaningful sequence. Assume that all facts and data presented in these notes are relevant; however, you may add or delete any facts or data to enhance the impact of your article. Also, you may interpret these facts and data in any manner that is consistent with your intent for the article.

When possible, include a technical description of the project, a phase of the project, or a process. Include, and cite in your text, simulated photographs with captions. Select an interesting title. Remember that the article must be interesting as well as informative.

a. Gridlock

Highways and flyways at a standstill

Companies losing money
  Documents not delivered on time
  Sales representatives late for presentations
  Private jets in holding patterns

Congestion will increase

Hydrocarbons pollute the air

Wasted fuel

  Lost productivity

Airline passenger travel has doubled in last decade

Motor vehicle travel has increased 27 percent in last decade

92 percent of U.S. roads built before 1960—almost 4 million miles

Decreased spending on public works from 19.1 percent of government expenditures in 1950 to 6.8 percent in 1986

Double-decker freeways

Vertical takeoff airplanes

300-mph magnetic levitation trains

43,000-mile interstate freeway system completed in 1991

$792 billion to move goods and people last year

4 percent of annual gasoline usage spent waiting at stop lights

Commuters show up to work tired and irritable

Freeway shootings

Recruitment of new employees difficult in gridlocked cities

Similar problems in Western Europe

Tokyo traffic increased by 49 percent—roads expanded by 4 percent

Women included in the work force

Teenagers with cars

Interstate system stimulated urban sprawl

Infrastructure needs repair—roads and bridges

Rail mass transit not effectively used by commuters

Encourage mass transit use by commuters

Residential neighbors prohibit night flights

More runways needed

Runway turnoff lanes during waiting periods

Military airfields for civilian use

b. Northridge Earthquake

January 17, 1994, 4:30 A.M. P.S.T.

6.6 on the Richter scale

61 deaths, 9000 injuries, $30 billion in damages

Northridge section of San Fernando Valley, 20 miles northwest of Los Angeles

2500 aftershocks in 10 days measuring 5.0 or greater

Earthquake located on unmapped fault, felt from San Diego to Las Vegas

6 major bridge collapses over freeways, 4 built before 1971 and scheduled for retrofit

Broke natural gas and water pipelines

Damaged or destroyed 45,000 residences

Unusual vertical thrust or acceleration

Most costly natural disaster in U.S. history

Team inspecting roadways and lifelines by January 19

Caltrans: ongoing seismic retrofit program since 1989

Interstate 10/405 interchange (retrofitted) in West Los Angeles virtually undamaged

I-10, I-5/14 interchange, Route 188 bridge collapses

Horizontal movement up to 10 feet

Vertical acceleration almost equal to horizontal acceleration, much higher than recorded before

Demolition of collapsed structures began hours after the quake, temporary structures in place

Cal State University at Northridge at center of the quake, 2-year-old $15 million parking structure destroyed

Steel buildings did well, but severe interior damage to equipment, elevators, utilities

Shopping centers—severe damage

Olive View Hospital—Sylmar did well, collapsed in 1971 quake, 6.4 on Richter scale

USC University Hospital with base isolators, minimal damage—0.5 g

Water system devastated—100 breaks including 7-ft-diameter pipe

Gas transmission lines did well

c. Nuclear Waste Disposal Site—Politics

Low-level radioactive waste site in Illinois

10 to 15 million cu ft of waste

Generated by nuclear power plants, research and medical laboratories, hospitals, pharmaceutical companies, industry

Steel and concrete vaults, 50-yr operating life, 560-yr design life

Illinois Department of Nuclear Safety (IDNS) selected Martinsville

Start-up expected by mid-1996

By mid-1995, under scrutiny for its political appeal

Site selected before technical confirmation

Independent Illinois Low-Level Radioactive Waste Disposal Facility

Siting Commission formed for approving proposed sites, 3-person Commission held 72 days of public hearings, 107 witnesses

October 1992, commission rejected Martinsville

1980 federal act makes states responsible for waste generated within their borders

1983—Illinois developed plan for storing, treating, and disposing of low-level radioactive waste

Martinsville—small rural town, facility would provide 100s of jobs

Initially, IDNS identified 21 possible counties for a site

Contractor criteria—4 sq mi minimum area and not contain an interstate highway

5 county boards voted to oppose siting—October 1987

4 sq mi criteria abandoned by contractor

Martinsville City Council endorsed the county as a site

Martinsville bisected by interstate highway

Questionable geology—Vandalia sand conveys contaminated ground water

Site surrounded by water, upstream of the city

Uncertain reinforced concrete would last 500 years

Members of commission resigned

IDNS now has only regulatory, not site-selection, responsibility

d. Towers With Foundation Mats

Four towers, Raffles City

$400 million hotel, office, and convention center

Singapore's waterfront

World's tallest hotel

World's largest monolithic concrete mat beneath

741-ft hotel, 125,000 tons

Differential settlement problems—different weights of towers

4.5 million ft^2

Designed by I. M. Pei and Partners, New York City

Built by Raffles City Pte., Ltd., Singapore

73-story hotel

42-story office tower

Twin 28-story hotel

Mat foundations, Dames & Moore, Los Angeles

Wet marine clay with sandstone boulders

Deep-bored piles were too expensive

Mats at different levels

Joints between mats capable of vertical movement

Started 3 years ago

484,000 cu yd excavation

Blasting of excavation not allowed

Dewatering system required

Storm channel rerouted

Mats—continuous pours of reinforced concrete

Plasticizers prevented concrete from setting too fast in hot climate

Towers—reinforced concrete—very economical

Site bounded by four heavily traveled roads

Weekend work—light traffic

Directional boom crane—one floor added every six days

Steel framing—top floors of 73-story tower
  30-ft high ceiling in restaurant
  Tall windows—1 inch thick, 700 pounds—installation difficult

Hotel and office buildings structure finished 3 months early

Delays on interior—wall panels, bathroom fixtures, tiles

e. Roller Coaster Technology

Safety a prime concern

$4 billion-a-year amusement park industry

Multimillion-dollar design and construction for new coasters

100-year history

Converts potential energy to kinetic energy

Wood-track coasters cannot turn you upside down like steel-track
  coasters

Shockwave at Six Flags Great America, Gurnee, IL, has seven loops

Round loops subject passengers to 12 g—fighter pilots pass out at 8 g

Oval or teardrop shapes reduce g to acceptable levels

5-inch-diameter welded steel pipe tracks

Suspended coasters in vogue—Iron Dragon at Cedar Point,
  Sandusky, OH—give an airborne illusion

Stand-up coasters in Tokyo—shoulder loops, tummy bar, and bicycle seat

Negative g are very controversial—passengers should not be lifted
  from their seats

Noise of wood tracks adds to the thrill for many riders

Wood coaster needs considerable maintenance

Illusion of danger important without real danger

20 fatalities from 1973 to 1985

Pairs of wheels ride above and below tracks for safety

Block systems: programmable controllers prevent collisions

Computer readout tells operator when seat restraints are locked

Passenger entry and exit are perilous

Programmable controllers determine coaster speeds

Brakes are metal-on-metal caliper types or metal brake fins combined paired with a composite lining

Bearings are usually lubricated with molybdenum disulfide

Daily safety inspections, yearly X-ray of critical metal parts

f. Japan's Monuments—21st Century

Innovative technology: industrial restructuring, advanced communications, infrastructure

Advanced lifestyle

Rainbow Bridge
Tokyo Bay: white, green, and coral lights
Opened August 1993—central Tokyo to waterfront city
570-m suspension bridge
6.5 years to build, $1.23 million
50-m clearance over 500-m-wide channel
Haneda International Airport—9 km away
Metropolitan Expressway Public Corporation
Environmental and visual impact
Water depth—12 m
Subsurface—alluvial stratum
762-mm-round cables—127 strands per cable

Landmark tower
70 stories—296 m, Yokohama's tallest
Opened June 1993, $2.7 billion
Waterfront development area
International business, culture, commerce, convention facilities
24 hours a day
Offices occupy first 49 floors; hotel, top floors
Corners incline inward—pyramid
Seismic, wind loads
Stubbins and Stubbins Architects of Boston, design
26-member joint venture construction

Haneda Airport Expansion
Land reclamation
Offshore development
Three-stage program: soil improvement, terminal building, new runways

Trans-Tokyo Bay Highway
Undersea tunnel, two man-made islands, bridge
15.1 km across Tokyo Bay
Kawasaki to Kisarazu
Studied for 8 years—technology from oil industry
Completion in 1996, $11 billion

CHAPTER 21

Experimental and Laboratory Test Reports

The results of procedures are communicated in laboratory reports.

Laboratory procedures determine the scientific and technical behavioral characteristics of the natural sciences and of man-made products and systems. These characteristics are described in laboratory reports.

The laboratory procedures used to determine these characteristics are performed by either the organizations that need this information or by independent laboratories that report the results to these organizations.

PURPOSE AND AUDIENCE

Laboratory procedures are typically used in professional practice in the following situations:

- In research, and research and development (see Chapter 23): Open-ended procedures (procedures without specific goals or tasks to be accomplished) are used to understand a theoretical concept such as the effect of temperature on a chemical reaction or a technical phenomena such as the mechanical properties of a new material. These procedures are typically fragility procedures (see Chapter 16) whose results are reported in experimental laboratory reports.

- For contractual or government regulatory compliance: These procedures are followed until a specific goal or task is accomplished to demonstrate that a product complies with an organization's specification, such as to demonstrate the capability of a vehicle's brakes to bring this vehicle to rest from a speed of 60 miles per hour within the maximum distance specified, or with a government regulatory code, such as to demonstrate compliance of an air exhaust system with the air quality regulations of the Environmental Protection Agency. These procedures are typically qualification procedures (see Chapter 16) whose results are reported in test (sometimes called qualification) laboratory reports.

- To obtain exploratory or deterministic data, which are used as the basis for further action: These procedures can be either experimental, such as when geotechnical laboratories produce borings to determine soil composition to design a foundation of a building, or test, such as to determine whether the cooling fins on a motor are adequate to prevent this motor from overheating.

An independent laboratory that performs a procedure for another organization (client) is responsible for complying with the client's procedural requirements and reporting the results of the procedure. A certificate of compliance (see the following section) certifies that the client's procedural requirements were met. However, the independent laboratory does not take any responsibility in the event that the procedure does not accomplish what was hoped for by its client.

Usually, an independent laboratory is required to submit its procedural specification (see Chapter 16) to its client for approval before beginning the procedure. This gives the client the opportunity to review the procedure, which may eliminate potential disputes on submittal of the final report.

The degree of formality of the laboratory report depends on its purpose and the intended readers. A formal report is usually required when the report is intended either to be used for further research, research and development, or to demonstrate compliance with a contractual obligation when the procedure is new or experimental. See the sample of a formal laboratory report in Chapter 11. The content of formal reports is discussed in this chapter. An informal report (which usually includes the name of the procedures, a data sheet, and calculations only) is sometimes used for routine fragility and test procedures or for fragility procedures that are part of an ongoing research or research and development program. See Figure 21–1 for a sample informal report.

CONTENT

Formal laboratory reports usually consist of the items in the following list. See Chapter 11 and Figure 21–2 for further discussion of these reports.

1. *Certificate of compliance:* a statement certifying compliance with the organization's procedural requirements. A certificate of compliance is not intended to imply that favorable data or results were obtained. Also, because the certificate is issued by the independent laboratory at the time that the report is submitted, it does not imply approval by the client to whom the report is submitted. Many independent laboratories that regularly issue certificates of compliance use standard forms for these certificates. See Figure 21–3 for a sample certificate of compliance.

Form 65. Results of Air-Compressor Test

Object of test _____

Owner _____ Builder _____

Test by _____ Location _____

Type of drive _____ Method of volume control _____

Cylinder diam _____ Stroke _____ Rated displacement _____ cfm

Rated speed _____ Barometric pressure _____

Item No.	Item	Units	Run No.	
			1	2
1	Duration of run	min		
2	Compressor discharge pressure gage	psi		
3	Temperature inlet air	°F		
4	Temperature of discharge air	°F		
5	Speed	rpm		
6	Kind and size of orifice			
7	Orifice pressure	in. water		
8	Orifice temperature	°F		
9	Indicated horsepower, air	ihp		
10	Temperature cooling water in	°F		
11	Temperature cooling water out	°F		
12	Weight of cooling water	lb/min		
13	Average piston speed	fpm		
14	Piston displacement per minute	cu ft		
15	Free air delivered per minute	cu ft		
16	Slippage	per cent		
17	Isothermal air horsepower	hp		
18	Adiabatic air horsepower	hp		
19	Gross horsepower	hp		
20	Volumetric efficiency	per cent		
21	Mechanical efficiency	per cent		
22	Compression efficiency	per cent		
23	Over-all efficiency	per cent		
24	Indicated horsepower in compressor cylinder per 100 cu ft of air delivered per minute at intake pressure and temperature	hp		

Date _____ Observers _____

_____

FIGURE 21–1

Sample Informal Laboratory Report

FIGURE 21–2
Sections of a Laboratory Report

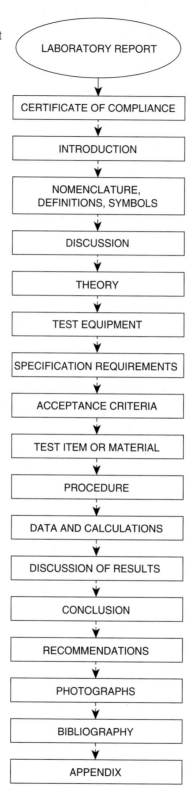

LABORATORY REPORT

CERTIFICATE OF COMPLIANCE

INTRODUCTION

NOMENCLATURE,
DEFINITIONS, SYMBOLS

DISCUSSION

THEORY

TEST EQUIPMENT

SPECIFICATION REQUIREMENTS

ACCEPTANCE CRITERIA

TEST ITEM OR MATERIAL

PROCEDURE

DATA AND CALCULATIONS

DISCUSSION OF RESULTS

CONCLUSION

RECOMMENDATIONS

PHOTOGRAPHS

BIBLIOGRAPHY

APPENDIX

CERTIFICATE OF COMPLIANCE

UNIT IDENTIFICATION NUMBER: QA–1203

SPECIFICATION NUMBER: 9211–A–8.0, Rev. 1 dated
March 25, 1995

ROTATION SPEEDS: 1800 Hz, 3600 Hz

All components were tested per Carr Company Test Procedure
L–82610 Rev. B, and the results of these tests are recorded as shown.

_____   _____
Mark Van Dam, P.E.   Date

Unico Laboratories	Test Certification	Date: 12/22/95	Pg. 1.1

FIGURE 21–3
Certificate of Compliance

2. *Introduction:* purpose of the procedure.

3. *Nomenclature, definitions, symbols:* any items of special meaning.

4. *Discussion:* applicability of the test method for obtaining the desired results, when not obvious.

5. *Theory:* development of any theory necessary to understand the procedure, data, or calculations.

6. *Test equipment:* a list or description of the test equipment used to perform the procedure.

7. *Specification requirements:* what the procedure is hoped to demonstrate. This is a summary of the design parameters and technical requirements as applicable to the procedure. This section is included in qualification reports only.

8. *Acceptance criteria*: the acceptable limits of damage or changes, as a result of the procedure, to the test specimen. This section is included in qualification reports only.

9. *Test item or material:* a description or identification of the test specimen.

10. *Procedure:* the step-by-step laboratory procedure. Sometimes, this procedure is included in the appendix. Unlike student laboratory reports, process descriptions are not used in professional laboratory reports.

11. *Data and calculations:* the numerical quantities obtained from performing the procedure, and the analyses of these data. Show graphs when appropriate.

12. *Discussion of results:* a discussion of the procedure, data, and calculations. Any procedural revisions that were required to successfully complete the procedure, and deviations from the anticipated results should be discussed.

13. *Conclusion:* what was determined by performing this procedure.

14. *Recommendations:* practical applications based on the conclusions from this procedure.

15. *Photographs:* the test specimen, equipment, test in progress, etc. to help the reader visualize the procedure. In qualification reports, include photographs of any damage or changes to the test specimen that occur as a result of the procedure.

16. *Bibliography:* a list of sources of material for further research.

17. *Appendix:* supplementary material, copies of original data sheets, etc. may be included.

KEY CONCEPTS

- Laboratory procedures are the bases for advances in science and technology. Their importance cannot be overemphasized.

- The laboratory reports that you write for your science and engineering laboratory classes are very similar to, and therefore are an excellent preparation for, writing professional laboratory reports.

STUDENT ASSIGNMENT

1. Submit a formal report to your technical writing instructor for a procedure that you performed for one of your science or engineering laboratory classes. You may deviate from the format recommended in this chapter (shown in Figure 21–2) so that you may meet the requirements of the laboratory instructor for whom this procedure was performed.

   Include a certificate of compliance stating that you complied with the procedural requirements of the laboratory class, and include a step-by-step procedure (rather than a process description) with this formal report.

   Submit a clean copy (i.e., without laboratory instructor comments) of this report to your technical writing instructor. Although the accuracy of your data will not be part of your grade, your presentation of this report, and your ability to technically express yourself coherently and logically—especially in your discussion of results, conclusions, and recommendations—will form the basis for your grade.

CHAPTER 22

Scientific and Engineering Analyses

Scientific and engineering behavioral characteristics are predicted and demonstrated in analytical reports.

Scientific and engineering analyses predict scientific and engineering behavioral characteristics and demonstrate that designs comply with appropriate codes, standards, and specifications. These analyses are the precursor to creative scientific and engineering achievement in research, development, and design.

PURPOSE AND AUDIENCE

Scientific and engineering analyses can accomplish the same results as laboratory procedures, but in significantly less time, and at considerably less expense. However, because assumptions are typically made to simplify these analyses, the results are significantly less reliable than laboratory results. Therefore, scientific and engineering behavioral characteristics of the natural sciences and man-made products and systems are frequently predicted by analysis before they are validated by performing laboratory procedures. However, after these behavioral characteristics predicted by analyses have been validated through laboratory procedures, these methods of analyses can be used for routine determinations of behavioral characteristics.

Similar to experimental and test laboratory procedures, scientific and engineering analyses are used in research and in research and development to understand a theoretical concept or technical phenomenon and to demonstrate contractual compliance of a product with a code, standard, or specification. Analyses are also used as part of the regulatory process of government agencies. Organizations usually perform these analyses internally, but consultants with expertise sometimes perform these analyses for clients.

Analyses that are intended to be submitted for approval are usually presented in the format of formal reports discussed in this chapter. For routine analyses, these formal reports are occasionally simplified into standard formats. Preliminary analyses intended to be studied further are frequently informal, because they are not for approval by others and sometimes include only the given data and calculations.

CONTENT

Because scientific and engineering analyses use mathematics and science as a base, these reports must be organized so that each section is developed from information presented in previous sections.

Theory and Calculations

Theory and calculations in a scientific or engineering analysis must be completely shown. In preparing these sections of the report, details are presented in a manner that a professional familiar with the discipline of the analysis can understand easily. Explanations precede all steps of the theory and calculations. The theoretical equations are shown immediately before introducing numbers into these equations. After the numbers have been introduced into these equations, intermediate mathematical steps used to solve them are not usually shown.

When a computer program is used for the solution of a system of equations, the methodology of this program should be explained before introducing this solution. When a spreadsheet is used to perform repetitive calculations, a sample calculation should precede this spreadsheet.

The theory and calculations shown in an analysis are unique inasmuch as text, graphics, theoretical equations, and calculations are integrated and presented in sequence to demonstrate the results. This facilitates the presentation for writers.

Graphics

Sketches show conceptual phenomena (e.g., free-body diagrams), physical phenomena (e.g., forces), and geometry. When pertinent, magnitudes of phenomena and dimensions are shown on these sketches. The critical parts of the sketches are labeled to help readers understand their meaning.

Tables with column and row headings that can be easily understood show sets of facts and data. The results section of a report usually includes a sketch showing the critical facts and data determined for the design. All graphics (e.g., sketches, tables) are introduced in the text with statements identifying them. When sketches are routinely integrated and presented in sequence with the text, for simplicity they are usually not labeled as figures.

Before release of a scientific or engineering analysis, the report should be independently reviewed for accuracy by another professional of the releasing organization.

When sketches and calculations are drawn by hand rather than with a computer, they should be of drafting quality, should be completed in pencil to facilitate changes, and should be written on vellum paper to obtain the most legible copies.

Multiple Design Conditions or Design Alternatives

When multiple design conditions (e.g., multiple locations of a moving load) or design alternatives (e.g., different possible geometric configurations) are considered, each condition or alternative must be individually shown and analyzed. Tables and spreadsheets are very useful for this situation. Calculations can be simplified for symmetrical conditions by including a statement to that effect and showing a sketch with critical facts and data determined by inspection.

Formal analytical reports typically include the following (Figure 22–1):

1. *Certificate of compliance:* a statement by the professional certifying compliance of the design and analysis with applicable codes, standards, and specifications. Often, a summary of the analysis and results are included in this certificate. See Figure 22–2 for a sample.

2. *Description of structure, system, or phenomena:* configuration, performance characteristics, and material properties.

3. *Summary:* the final determination based on the calculations. This is usually in the format of a sketch. This section is frequently placed at the beginning of the report for the convenience of the reader.

4. *Specification requirements:* design criteria, loads and special requirements used as a basis for the analysis.

5. *Scope:* a statement of the structure, system, or phenomena analyzed and determined in the report. Items not considered in the analysis should be addressed in this section.

6. *Theory and assumptions:* development of theory and the assumptions for that theory used as a basis for the analysis.

7. *Symbols, nomenclature:* a list of items with special meanings.

8. *Analysis:* the calculations and its results.

9. *Conclusions:* the conclusions based on the analysis.

10. *Design alternatives considered:* alternatives considered as a consequence of the analysis.

Sample Scope Written for This Assignment

In this design and analysis, it is assumed that the weight of the bridge is negligible and the weight of the welds and joints is 10 percent of the member weights. It is also assumed that all forces in the structural members are axial. Stenium alloy, a buckle-resistant material, is used for the design of the members.

Because the bridge is symmetrical, member loads for Case 3 are determined by inspection by comparison with Case 1.

Critique of Sample Scope

Although the information that is presented in this section is technically correct, the emphasis is on the assumptions for the analysis rather than on the structure

FIGURE 22–1

Sections of an Analysis

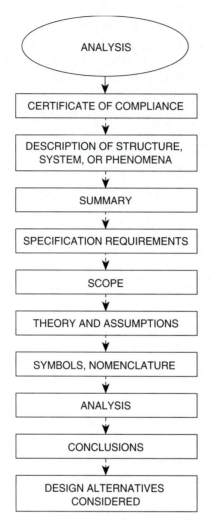

that is designed and analyzed in the report. The first paragraph is merely a list of assumptions and material properties. The second paragraph relates to the method of analysis.

The purpose of the scope section of the report is to inform the reader of the work that is intended to be accomplished—in this case, the design and analysis of the two side trusses of a bridge.

The weight of the bridge is assumed to be negligible for designing the trusses. However, this assumption limits the validity of the design of the bridge rather than states the work that was accomplished. Although this assumption is critical to the validity of the design, it does not address what was designed and analyzed and therefore is not appropriate to be addressed in this section.

CERTIFICATE OF SEISMIC COMPLIANCE

CUSTOMER: Sugaree Services
UNIT IDENTIFICATION NUMBER: AR–16
SPECIFICATION NUMBER, DATE: LN–1521 Rev. 2 dated Dec. 18, 1994
CRITICAL DESIGN CONDITION: Design Basis Earthquake
CRITICAL MINIMUM NATURAL FREQUENCY: 33 Hz
SEISMIC ACCELERATIONS: h = 1.00g, v = 0.71 g

SUMMARY

The above structure is analyzed for horizontal and vertical seismic loadings applied simultaneously with operating loads. All critical members and boundary supports are analyzed for the critical condition shown above and show a positive margin of safety. Natural frequencies of all structural support members are greater than the critical natural frequency. Deflections are investigated and are not significant.

CERTIFICATION

The reactor housing is structurally adequate to function for the most critical seismic conditions as required by the customer specifications.

_____   _____
Jerry Guzman P.E.   Date

GD Designs	Reactor Housing	Date: 6/14/95

FIGURE 22–2
Certificate of Compliance

A more appropriate scope might read as follows:

> The two side trusses are designed and analyzed, and the weight of these trusses is calculated. The approach to the bridge, the end-supports, the roadway, and the cross-members are not the subject of this design and analysis.

Notice that a 100,000-lb load that the bridge is designed to withstand is a design condition, not what is hoped to be accomplished, and is not addressed. The method of calculating the weight and the material properties of Stenium alloy is not part of what is hoped to be accomplished and is also not addressed.

SAMPLE ASSIGNMENT—BRIDGE DESIGN AND ANALYSIS

You are employed as a structural engineer by a structural engineering firm. Your client, a government agency, contracted your firm to design and analyze a bridge to be made of Stenium alloy. You are required by the government specification to use no more than three structural sizes (use rectangular shapes to the nearest square inch; i.e., do not select shapes from the AISC Handbook) for simplicity of fabrication. The sizes of the members are to be selected so that the composite weight and cost of the structure are minimized. The specification requires that 10 percent be added to the estimated weight of the structure to account for welds and joints.

In your design and analysis, include calculations to demonstrate that alternative structural sizes were considered for the reduction of weight in your design. Do not forget that the bridge consists of two trusses, and therefore each truss supports only 50 percent of the design load. For simplicity of this assignment, assume that the weight of the bridge is negligible to calculate the member loads. Of course, this assumption is unreasonable in professional practice, and design practice would require a reiterative analysis. Also remember that Stenium alloy has been developed very recently in the secret laboratory of your instructor, and its metallurgical and mechanical properties are confidential and therefore unknown to the engineering community. Your design is for the side trusses only and does not include the roadway, cross-members, end-supports, or the approach ramps.

You have received the letter shown in Figure 22–3 from your client confirming your contract to design and analyze a bridge.

ABC METROPOLITAN TRANSPORTATION DISTRICT
P.O. BOX 361
OURCITY, US 12345
(313) 421-4141

[date]

[your name]
Bridge Design, Inc.
P.O. Box 305
Yourcity, US 54321

Dear Mr./Ms. [your name]:

I enjoyed our lunch meeting last week and discussion of our new project.

This letter confirms that you will commence the design and analysis of the truss structure of the 80-foot bridge to span the Kiwi River. The configuration of the bridge and the material properties of Stenium alloy, a new buckle-resistant material, are shown in the attachment.

The bridge must be capable of supporting 100 kips (1 kip = 1000 lb) from any joint along the roadway. The weight of the bridge may be assumed to be negligible for the design and analysis. The area of each structural member is to be optimized; however, no more than three structural sizes may be used for ease of fabrication. Each structural size will be rectangular and is to be specified to the nearest square inch. Please include a composite weight of the bridge with your analysis including a 10 percent allowance for welds and joints so that we can estimate the cost of materials and fabrication.

The final design and analysis is due on [date]. Enclosed is a $1000 retainer to begin work. Please submit your invoice at $70per hour plus expenses with the report.

Please call if you have any questions.

Very truly yours,

Les Werke

Les Werke, P.E.

Attachment

FIGURE 22–3
Sample Client Confirmation of Contract for Scientific Design and Analysis

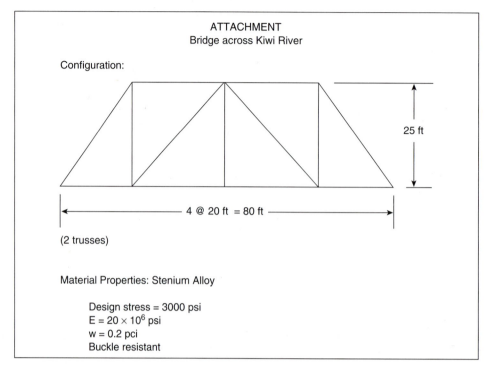

ATTACHMENT
Bridge across Kiwi River

Configuration:

25 ft

4 @ 20 ft = 80 ft

(2 trusses)

Material Properties: Stenium Alloy

Design stress = 3000 psi
$E = 20 \times 10^6$ psi
$w = 0.2$ pci
Buckle resistant

FIGURE 22–3
(*continued*)

Critique of Sample Design and Analysis

As shown in Figure 22–4, the cover letter, included with any formal report to a client, is presented well. Its content appropriately states what is enclosed, presents a short summary of what is accomplished, and has a pleasant gratuitous closing.

The cover page, certificate of compliance bound in the report behind the cover page, and table of contents are professional-looking.

The invoice, inserted loosely behind the title page, has a professional-looking appearance and includes all the information that the accountant may want to know before making payment. However, the balance due is obviously incorrect. This error probably resulted from the student using an invoice from a previous assignment as a template and making the necessary changes. A simple error of this nature can cloud the credibility of your calculations in the report; therefore, great care should be taken to prevent these errors.

A summary of what is accomplished should be included at the beginning of the report. This could best be done with a sketch of the truss showing the member sizes determined and the composite weight of the structure. This result is not summarized anywhere in this report.

The project description should include the sketch shown on the top of page 2.

The scope includes what is intended to be accomplished, but similar to the sample scope shown previously, it erroneously includes assumptions, method of analysis, and material properties.

The sketches in the analysis are easy to understand, although the joints for Load Cases 1 and 2 shown on page 2 should be identified.

Notice that the diagrams in this report are labeled but do not need figure numbers because they are an integral part of the text and calculations.

To simplify reviewing the analysis, the free-body sketches and the equations beginning on page 3 should always include numerical values when they have been determined, rather than showing only their letter designations.

The summary of member forces table is an effective summary of all load conditions used to determine the critical condition for the design. Note that the column headings are brief yet sufficiently specific. A sketch summarizing the critical loads shown in the table should be included.

A statement is needed preceding the design area table to introduce the data. An introductory statement is also needed before the member weights table.

Greater attention needs to be brought to the final member area selections and the composite weight shown in the member weights table. This can be accomplished with a bolder type, underlines, arrows, or other mechanical devices.

Although the calculations in this report are correct, the appendix includes an incorrect design stress. This is probably a transcription error, which should have been corrected.

KIWI RIVER BRIDGE

DESIGN AND STRUCTURAL ANALYSIS

Job No. 95–137

Date Submitted: November 20, 1995

Prepared by:

Bridge Design, Inc.
Yourcity, US

Prepared for:

ABC Metropolitan Transportation District
Pomona, CA

FIGURE 22–4

Sample Design and Analysis (pp. 336–348)

BRIDGE DESIGN, INC.
P. O. Box 305
Yourcity, US 54321
(714) 713–1313
(714) 713–1377 fax

November 20, 1995

Les Werke, P. E.
ABC Metropolitan Transportation District
P. O. Box 361
Ourcity, US 12345

Dear Mr. Werke:

Enclosed is the structural design and analysis of the proposed Kiwi River Bridge. A current invoice for engineering services performed on this project and a certificate of compliance are also included.

The size and configuration of the bridge complies with the requirements of the specification of your company. The 80-ft. bridge is designed to support 100 kips from any joint along the roadway using the given material properties. The structural sizes of Stenium alloy members were analyzed for weight optimization.

I hope you are pleased with the design and structural analysis. I appreciate the opportunity to work with you and look forward to working with you again.

Please call me if you have any questions.

Sincerely,

Marlene E. Kew

Marlene E. Kew, P. E.
Associate Engineer

Encl: Kiwi River Bridge Design and Structural Analysis
        Invoice
        Certificate of Compliance

BRIDGE DESIGN, INC.
P. O. Box 305
Yourcity, US 54321
(714) 713–1313
(714) 713–1377 fax

INVOICE

DATE: November 20, 1995
CLIENT: ABC Metropolitan Transportation District
JOB NO: 95–137
Kiwi River Bridge Design and Structural Analysis

Engineering Time

Design, Oct. 20, 1995 thru Nov. 17, 1995	8.0
Structural Analysis, Oct. 25, 1995 thru Nov. 10, 1995	20.0
TOTAL	**28.0 hr**

Engineering Services and Expenses

Engineering (28.0 hr @ $70/hr)	$1,960.00
Structure Code Book	25.00
TOTAL	**$1,985.00**
Less Retainer	<$1,000.00>
BALANCE DUE	**$4,310.00**

Note the error in the balance due, which should be $985.00, not $4310.00

FIGURE 22–4
(*continued*)

BRIDGE DESIGN, INC.
P. O. Box 305
Yourcity, US 54321
(714) 713–1313
(714) 713–1377 fax

November 20, 1995

ABC Metropolitan Transportation District
P. O. Box 361
Ourcity, US 12345

CERTIFICATE OF COMPLIANCE

Project: Kiwi River Bridge Design and Structural Analysis
          Job No. 95–137

I hereby certify that the structural design and analysis of the Kiwi River
Bridge complies with all governing state and local standards and is in
accordance with the requirements specified by ABC Metropolitan
Transportation District.

Sincerely,

Marlene E. Kew

Marlene E. Kew, R. C. E. 77137

This report should include a summary of results with the final member sizes and weight.

TABLE OF CONTENTS

FIGURE 22–4
(*continued*)

KIWI RIVER BRIDGE

1.0 Introduction

This report, which includes the design and analysis of Kiwi River Bridge, is prepared by Bridge Design, Inc. This design and analysis complies with the bridge design standards and structural analysis standards provided by ABC Metropolitan Transportation District.

2.0 Project Description

The proposed Kiwi River Bridge spans 80 ft across the Kiwi River. A twin-truss infrastructure will support the roadway with one truss on either side of the roadway. The two trusses will each be 25 ft in height and fabricated from Stenium alloy (material properties are given in the Appendix).

3.0 Scope

Assumptions, method of analysis, and material properties should be included elsewhere in the report. The scope should include only what is intended to be accomplished.

The twin-truss structure of the bridge analysis is designed to support a load of 100 kips, or 50,000 lb per truss, at any joint along the roadway. For the design and analysis, forces in each member of the truss are assumed to be axial and the weight of the bridge is assumed to be negligible. A composite weight of the bridge, however, will be included with a 10 percent allowance for welds and joints. The design of the roadway and cross-members is not the subject of this report. The moments of inertia of the individual members are not calculated, due to the unique buckle-resistant properties of Stenium alloy.

A bridge design is determined that most efficiently uses a maximum of three different-sized structural sections.

1

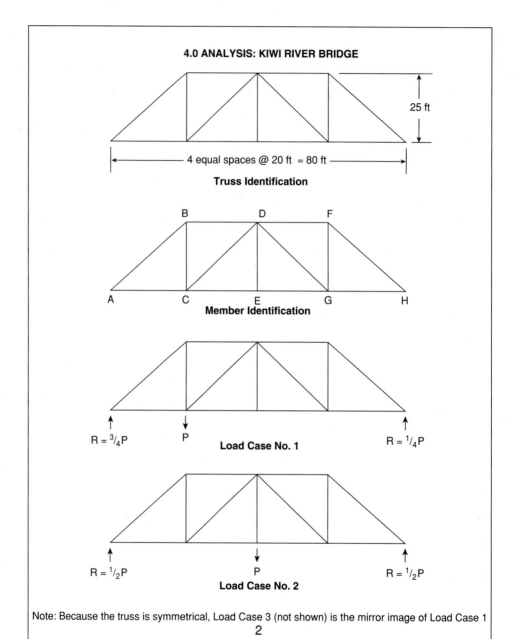

FIGURE 22–4
(*continued*)

Load = 100,000 lbs

$$P = \frac{100,000 \text{ lbs}}{2 \text{ Trusses}} = 50,000 \text{ lb/Truss}$$

METHOD OF JOINTS

LOAD CASE 1

JOINT A

$\Sigma F_X = 0$ $- AB \frac{20}{\sqrt{1025}} + AC = 0$

$\Sigma F_Y = 0$ $- AB \frac{25}{\sqrt{1025}} + R_A = 0$

$AB = \underline{48,024 \text{ lb}}$
$AC = \underline{30,000 \text{ lb}}$

$R_A = 37,500\#$

JOINT B

$\Sigma F_X = 0$ $- BD + 48,024 \frac{20}{\sqrt{1025}} = 0$

$\Sigma F_Y = 0$ $- BC + 48,024 \frac{25}{\sqrt{1025}} = 0$

$BD = \underline{30,000 \text{ lb}}$
$BC = \underline{37,500 \text{ lb}}$

JOINT C

$\Sigma F_X = 0$ $- AC + CE + CD \frac{20}{\sqrt{1025}} = 0$

$\Sigma F_Y = 0$ $BD - P + CD \frac{25}{\sqrt{1025}} = 0$

$CD = \underline{16,008 \text{ lb}}$
$CE = \underline{20,000 \text{ lb}}$

$P = 50,000\#$

JOINT E

$\Sigma F_X = 0$ $- CE + EG = 0$

$\Sigma F_Y = 0$ $DE = 0$

$EG = \underline{20,000 \text{ lb}}$

3

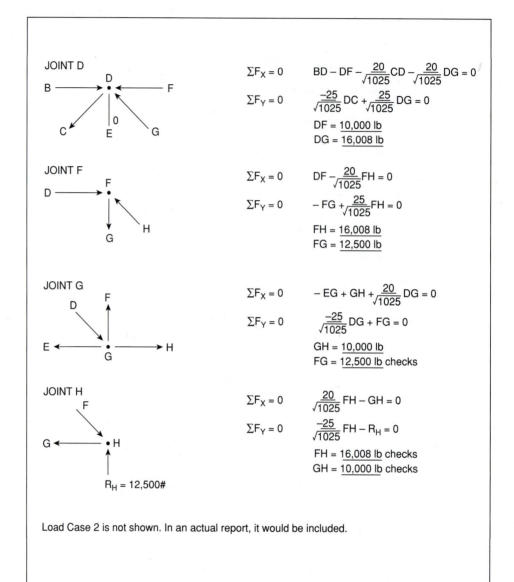

FIGURE 22–4
(*continued*)

Summary of Member Forces

The forces in each member are:

Member	Case 1 (lb)	Case 2 (lb)	Case 3 (lb)	Critical Design Force (lb)	Length (ft)
AB	48,024	32,015	16,008	48,024	32.02
AC	30,000	20,000	10,000	30,000	20.00
BC	37,500	25,000	12,500	37,500	25.00
BD	30,000	20,000	10,000	30,000	20.00
CD	16,008	32,015	16,008	32,015	32.02
CE	20,000	40,000	20,000	40,000	20.00
DE	0	50,000	0	50,000	25.00
DF	10,000	20,000	30,000	30,000	20.00
DG	16,008	32,015	16,008	32,015	32.02
EG	20,000	40,000	20,000	40,000	20.00
FG	12,500	25,000	37,500	37,500	25.00
FH	16,008	32,015	48,024	48,024	32.02
GH	10,000	20,000	30,000	30,000	20.00

5

An introductory statement is needed to explain the data in the table.

Design Area

Member	Minimum Area (in.^{2})	Design 1 (in.^{2})	Design 2 (in.^{2})	Design 3 (in.^{2})	Design 4 (in.^{2})	Design 5 (in.^{2})
AB	16	17	16	17	17	17
AC	10	10	13	12	11	10
BC	13	14	13	15	14	17
BD	10	10	13	12	11	10
CD	11	14	13	12	11	12
CE	14	14	16	15	14	17
DE	17	17	17	17	17	17
DF	10	10	13	12	11	10
DG	11	14	13	12	11	12
EG	14	14	16	15	14	17
FG	13	14	13	15	14	17
FH	16	17	16	17	17	17
GH	10	10	13	12	11	12

6

FIGURE 22–4
(*continued*)

An introductory
statement
explaining the
data should be
included before
the title.

Member Weights

Member	Design 1 (lb)	Design 2 (lb)	Design 3 (lb)	Design 4 (lb)	Design 5 (lb)
AB	1,306	1,230	1,306	1,306	1,306
AC	480	624	576	528	480
BC	840	780	900	840	1,020
BD	480	624	576	528	480
CD	1,076	999	922	845	922
CE	672	768	720	672	816
DE	1,020	1,020	1,020	1,020	1,020
DF	480	624	576	528	480
DG	1,076	999	922	845	922
EG	672	768	720	672	816
FG	840	780	900	840	1,020
FH	1,306	1,230	1,306	1,306	1,306
GH	480	624	576	528	576
Total Wt	10,729	11,069	11,021	10,459	11,165

+10% for connections:

	11,801	12,176	12,123	11,505	12,282

Total weight for 2 trusses

	23,603*	24,352	24,247	23,011	24,563

*Select Design 4 with area members 17, 14, 11.

Sample Calculations

Design area = (design force) / (design stress)

Weight = (design area) (length) (w)

7

The design stress is incorrect. It should be 3000, not 2500.

5.0 APPENDIX

Stenium Alloy Properties

Design Stress = 2500 psi

Modulus of Elasticity = 20×10^6 psi

Unit Weight (w) = 0.2 1b/cf

Buckle Resistant

8

FIGURE 22–4
(*continued*)

KEY CONCEPTS

- Analysis is the basis for determining routine scientific and engineering behavioral characteristics when the method of analysis has been validated by results from laboratory procedures.

- Analytical reasoning to understand scientific and engineering behavioral characteristics is the precursor to creative achievement by professionals. Research, development, and design require this ability.

STUDENT ASSIGNMENT

Write a formal analytical report for one of the following:

1. Submit a formal analytical report to your technical writing instructor for a design project completed for one of your design-related classes or, a work-related project. You may deviate from the format recommended in this chapter (shown in Figure 22–1) to meet the requirements of your design instructor or employer.

   Include a certificate of compliance certifying compliance of the design and analysis with either the appropriate text assignment or, for a work-related project, the applicable codes, standards, and specifications. Also, include a summary on this certificate.

   Submit a clean copy (i.e., without instructor comments) of this report to your technical writing instructor. Although the accuracy of this report will not be part of your grade, your presentation of this report and your ability to express your results coherently and logically will form the basis for your grade.

Problems 2 and 3 require a knowledge of the principles of strength of materials.

2. Design a ladder to hold the maximum weight person using the following criteria:
   - This ladder shall consume 40 (or 60, 80, etc.) pounds of wood, which has a unit weight of 50 pounds per cubic foot.
   - This ladder shall be 6 (or 8, 10, etc.) feet long.
   - The rungs of this ladder shall be solid circular bars spaced 12 inches apart. Rungs shall not be included at the very bottom or very top of the ladder.
   - The siderails shall be solid square bars. The inside width between these siderails shall be 12 (or 14, 16, etc.) inches.
   - The design stress shall be 1000 (1200, 1500) psi.

   The design of the joints of the rungs to siderails will not be the subject of this report. For simplicity, ignore the weight of the ladder (this is unrealistic in design practice) in your calculation of stresses.

Assume that the critical design condition of the ladder occurs when it is used as a horizontal beam that spans two supports spaced 6 (or 8, 10, etc.) horizontal feet apart. Also, you may assume that buckling and deflection are not critical for this design.

To design this ladder to hold the maximum possible weight, all the wood that is available should be used. Because this wood must be efficiently distributed to the rungs and siderails, the solution will be trial-and-error. Use spreadsheets and tables when possible.

Assume that the readers of your report are engineers in another discipline who have a basic knowledge of engineering principles.

3. Design a statically determinate (e.g., a Pratt truss) bridge to hold the maximum single point load from any joint along the roadway using the following criteria:
   - This bridge shall be 100 (or 120, 140, etc.) feet long and consist of 4 panels.
   - The height of the bridge shall be 25 (or 30, 35, etc.) feet tall.
   - This design shall consume 150,000 (or 200,000; 250,000; etc.) cubic inches of steel.
   - All members shall be solid square bars.
   - The areas of compression members shall be twice the areas of tension members.
   - The design stress shall be 20,000 (or 25,000; 30,000; etc.) psi.

The designs of the joints and roadway for this bridge are not the subjects of this report.

For simplicity, ignore the weight of the bridge (this is unrealistic in design practice) in your calculation of stresses. Also, you may assume that buckling and deflection are not critical for this design. Do not forget that there are two parallel trusses for this bridge.

To design this bridge to hold the maximum possible load, all the steel that is available should be used. Because this steel must be efficiently distributed to the primary and secondary members, the solution will be trial-and-error. Use spreadsheets and tables when possible.

Assume that the readers of your report are engineers in another discipline who have a basic knowledge of engineering principles.

CHAPTER 23

Research Reports

*Advances in science and society are made possible through the
results of research.*

PURPOSE AND AUDIENCE

The purpose of research is to foster understanding of scientific and sociotechnologic (the effect of technology on society) concepts. Using known facts and data,
research predicts the unknown, such as the mechanical properties of an experimental material, why some birds fly higher than other birds, and the effect of
smog on productivity. The results of research are for information purposes only
and ordinarily are not intended to have an immediate application.

Research reports document these findings for use in future research and for the
development of applications. These reports may be used within either the
researching organization or the organization for which the research is performed,
or they may be published in technical or professional journals.

The findings of research can be based on a literature search of previous relevant investigations, extrapolation of data from previous laboratory procedures,
and the results of laboratory procedures (see Chapter 21) and scientific analyses
(see Chapter 22) performed by the researcher. To be most effective, researchers
typically determine their findings by combining these methods.

Research frequently begins with a literature search of reports, journals, and textbooks to determine the present available information. Particular attention is paid
to published test data. Scientists usually use this information as a basis to perform
scientific analyses and experimental laboratory procedures. Sociotechnologists
usually use this information as a basis for analyses and gathering additional relevant data and information, which may be obtained from data surveys, question-
naires, and sociotechnologic laboratory procedures (similar to scientific laboratory
procedures except that the object of the procedure is to determine the effect of
technology on society).

FUNDING FOR RESEARCH

Research projects are usually funded by one of the following:

- A government agency (federal, state, or local) that either performs its own research or contracts this research to a private organization by competitive bidding. Research by a government agency is performed as part of the research and development function of the government.

- A foundation or other nonprofit organization that either performs its own research or gives grants to other organizations (frequently other nonprofit organizations such as universities) to perform this research. These grants are obtained by the research organization by grant proposals (see Chapter 15).

- A private organization that performs its own research to remain technically competitive. This is common in the high-tech industries such as electronics and computers.

The source of funding for development and for research and development (both are discussed later in this chapter) can be any of the same sources as for research. However, because development is intended for practical application and to be profit-generating, these projects are frequently funded by private organizations.

RESEARCH AND DEVELOPMENT COMPARED TO RESEARCH

Research uses the techniques of information search, analysis, and laboratory procedures to determine *findings*. Research and Development (R&D) uses these same techniques to determine *practical applications*. The differences between R&D and research are their informational bases and their goals.

Research begins with a known body of scientific or sociotechnologic knowledge, which this research intends to expand. The goal of this expansion usually is to demonstrate that a hypothesis accurately predicts an outcome, and any results that are obtained are deemed acceptable, provided that the methodology to obtain these results is valid. These results may be applied science (information that may be used for scientific application) or applied sociotechnology.

By comparison, R&D begins with an applied science or an applied sociotechnology, which it develops into a practical application. Therefore, the goal of R&D is defined before the process begins. When this R&D process does not result in a practical application, it is revised and repeated until a desired result is obtained. Therefore, R&D may be an iterative procedure.

R&D includes the determination of qualitative and quantitative concepts for primary design concepts (e.g., size, process, location); analysis of these concepts; and when necessary, laboratory procedures to validate the feasibility of these concepts. Detail design (quantitative determination of all variables) is completed in the design phase subsequent to R&D. The result of detail design is a set of drawings that is used by technicians to build or manufacture.

The report that results from R&D ordinarily begins with the given body of knowledge and demonstrates the feasibility of the final application. Intermediate iterations used for determining the final application are not usually shown. Similarly, a research report ordinarily begins with a given body of knowledge and then demonstrates (or fails to demonstrate) that a hypothesis accurately predicts an outcome. Therefore, the format of an R&D report is similar to that of the research report discussed in this chapter.

DEVELOPMENT ONLY COMPARED TO R&D

Development is a process that begins with a known technology or known sociotechnology (a practical application of science or sociotechnology), which it uses to find other practical applications, whereas R&D begins with applied science or applied sociotechnology to find those same practical applications. However, the process and the goals of development and R&D are usually similar.

BRIDGING THE GAP FROM R&D TO DEVELOPMENT

The R&D process and the development process usually begin when a need is perceived rather than when research has discovered new findings. When this need arises, the challenge to the professional is to create a solution that satisfies this need using the existing bodies of knowledge. The ability to create is what distinguishes an intelligent being from a machine. This ability allows us to reach back from the perceived need into the existing body of knowledge to find a practical solution to meet this need.

Creating a Solution

Some people need to work alone to be creative, while others need to brainstorm in a group. Most people benefit by doing some of each.

The sequential phases of the creative process for most professionals include the following:

1. Research all possible relevant information.

2. Determine possible solutions to satisfy the need.

3. Relax the conscious mind with rest or recreation to allow the subconscious mind to review these possible solutions.

4. Determine the best possible solution from these subconscious thoughts by conscious evaluation.

   Steps 2 to 4 are usually reiterative until the professional is convinced that this solution is the most technically and economically feasible of all possible solutions.

5. Develop the technology for this solution.

CONTENT

Information in a research report may be used for further research or development. Therefore, present the information so that your research can be duplicated by the reader. Similar to laboratory and test reports (see Chapter 21), and analytical reports (see Chapter 22), the testing and analysis must include step-by-step details.

Graphics help readers understand the content and enable them to apply the results of the research.

The following sections are typically included in research reports (Figure 23–1):

1. *Abstract:* an informative abstract that summarizes the information in the report. It includes conclusions and recommendations.

2. *Introduction:* the function of the report, its purpose, and its source of authorization and funding.

3. *Scope:* a statement of the body of information researched in the report. Peripheral items not addressed in the report should be stated.

4. *Purpose:* the goals or hypothesis of the research.

5. *Symbols, nomenclature:* a list of items with special meanings.

6. *Given information:* a statement of the problem.

7. *Assumptions:* a list of facts assumed to be true.

8. *Research:* introduction of information based on the literature search, analyses, and laboratory procedures. See Chapters 21 and 22 for discussion of presenting laboratory procedures and analyses into reports.

9. *Results:* a summation of the findings based on the research.

10. *Conclusions:* the determinations drawn from the results.

11. *Recommendations:* a statement of what is recommended as a result of this research.

12. *Bibliography:* a list of sources of additional information. Footnotes can be used for explanations of the text material. They are placed at the bottom of each page to make it easy for the reader to use them. The numbering can either begin at each new page, or can be sequential throughout the report. The note number in the body of the text is superscripted and is placed at the end of the passage to which it pertains. Endnotes are similar to footnotes except they are placed at the end of the report and are numbered consecutively throughout the report.

Sample Introduction

Contractors and manufacturers must consider the potential problem of corrosion when steel is used as a structural material. The research in this report addresses the question, What causes steel to corrode?

FIGURE 23–1
Sections of a Research Report

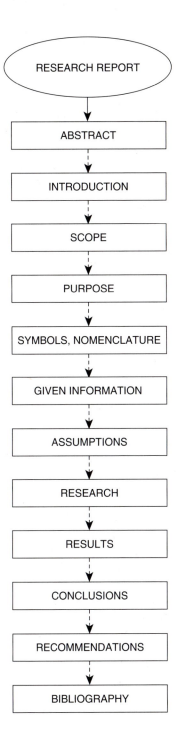

The research and report are funded through a grant of the National Steel Foundation and are in fulfillment of Contract 560-102 dated July 17, 1995.

Critique of Sample Introduction

This introduction addresses the purpose of the report, its source of funding, and its authorization. These facts are stated clearly and directly.

SAMPLE ASSIGNMENT—RESEARCH REPORT

Assume that our body of scientific knowledge is limited to the science and mathematics textbooks used for your introductory classes. Write a research paper to explain the causes of corrosion of steel.

Figure 23–2 (pp. 358–362) illustrates a typical research report on the subject of the assigment.

Critique of Sample Research Report

The sample research report shown in Figure 23–2 is an excellent example of the ability of a student to write an effective report. It is carefully organized, has multiple graphics, and has a professional appearance. However, sections may need to be revised before this report is submitted.

Because of its organization and presentation, any reader of this report would clearly understand the information included and have confidence in its results. When the quality and the presentation are logical and clear, and the quality of the research is clear, the releasing organization has a reason to be proud.

An abstract is noticeably missing at the front of the report.

The subject of the report is discussed in the introduction, but in a professional environment, the authorization for a research report and its source of funding also need to be identified.

The scope clearly defines the chemical composition of the samples and temperature ranges that are researched in the report. It also appropriately identifies similar steels that are not included in the research.

The purpose clearly identifies what is hoped to be determined from this research. Although this research has a specific goal, no hypothesis is indicated, because it is unknown in this case.

The test specimen includes the necessary data, but the information could be expressed more concisely.

The given information is stated as if it were intended to be a background to the problem. Rather, the given information could have read, "Joints of water pipes of old houses and roofs of barnyard structures develop a brown coating of unknown origin after 6 months of exposure to sun, rain, dew, light, darkness, smog, and temperatures ranging from 30°F to 120°F."

The research section is a series of laboratory procedures. A table of results for each factor is preceded by a description of that procedure. A conclusion follows each table of results. The tables are neatly presented. Columns and rows are clearly headed and include units. A summary table at the end of the research section makes the factors easy to compare.

The recommendations section appropriately recommends actions for preventing rust.

Because no references are included in the report to other sources of information that are necessary for understanding its contents, a reference section is not included. However, a bibliography including sources of information researched for this report also is not included. This implies research without the use of other sources, an unlikely event.

The date of sub-
mittal should be
included on the
cover page.

RESEARCH REPORT: THE CAUSES OF CORROSION OF STEEL

TABLE OF CONTENTS

[*For space considerations, these sections have been omitted from this text.]

FIGURE 23–2
Sample Research Report (pp. 358–362)

Introduction

Steel is the most useful metal known to humans. With a variety of uses, which range from paper clips to space vehicles, steel has become an important part of modern civilization. Unfortunately, steel has an inevitable problem: corrosion. As our product and manufacturing processes have become more complex, the penalties of structural failure caused by corrosion have become more costly and therefore better recognized. This report discusses the causes of corrosion.

An abstract should precede the introduction.

Scope

This report discusses medium steels (containing from 0.2 percent to 0.6 percent carbon), which are used for making rails and structural members such as beams, girders, etc.

Mild steels (low carbon steels with less than 0.2 percent carbon), high carbon steels (containing 0.75 percent to 1.50 percent carbon), and alloy steels (containing considerable amounts of metals other than iron) are not of concern.

This report discusses the corrosion of steel only at temperatures from 30° F to 120° F.

Purpose

The purpose of this research is to determine the factors that cause steel to corrode. Therefore, steel specimens were exposed to different environments: atmospheric (seacoast, industrial, rural), water (pure water, seawater, freshwater), and different temperatures.

Based on the results of this research, it was determined that the life of steel can be prolonged by preventing corrosion.

Test Specimens

Size and shape of test specimens: A standard size and shape was used for all specimens in each test series:

10-cm $\times$ 20-cm $\times$ .1-cm panel.

1

Definition

Corrosion rate: The ratio of the mass loss per 100 g to the time exposed (100 days).
$$R = \Delta m\, /t \qquad \text{(mm/day)}$$

Given Information

When looking at the water pipes of an old house, one can see the heavy corrosion at the joints. Barnyard roofs made of steel also have a brown coating on their surfaces.

Research

1. TEMPERATURE:

Ten steel specimens placed in different tubes ranging from 30° F to 120° F, for 100 days.

Tube No.	Temp. (° F)	Original Weight (g)	Final Weight (g)	Weight Loss (g)	Rate of Corrosion
1	30	350	349.76	.24	.69
2	40	350	349.63	.37	1.05
3	50	351	350.59	.41	1.17
4	60	350	349.44	.56	1.60
5	70	352	351.28	.72	2.05
6	80	351	350.11	.89	2.53
7	90	350	348.47	1.03	2.94
8	100	351	349.81	1.19	3.39
9	110	352	350.66	1.34	3.81
10	120	351	349.44	1.56	4.44

The increase in the rate of corrosion of steel demonstrates that temperature is a factor.

2

FIGURE 23–2
(*continued*)

2. WATER:
The specimens were submerged in pure water, seawater, and freshwater at 60° F for 100 days.

	Original Weight (g)	Final Weight (g)	Weight Loss (g)	Rate of Corrosion
Pure water	351	350.90	.1	.28
Seawater	350	347.21	2.79	7.97
Freshwater	350	349.10	.9	2.57

The rate of corrosion in steel was highest for seawater. This demonstrates that the salt in the water is a factor that causes steel to corrode.

[For space considerations, pages 4 through 6, containing the research concerning atmosphere, sunlight, and oxygen, has been omitted from this text.]

3

Conclusion

The following summarizes results when tests performed at 60° F for 100 days.

Specimen	Rate of Corrosion
Pure water	.28
Seawater	7.97
Freshwater	2.57
Industrial area	2.27
Non-industrial area	.56
Dark place	.11
Sunlight place	.14
Vacuum	0.00
Oxygen tank	12.54
60° F in tube	1.60
120° F in tube	4.44

No corrosion occurred in the vacuum tube, even at a high temperature. Pure water, without the solvent, does not cause undue corrosion of steel. The chemistry of the water (hardness, salts, chlorides, dissolved gases) and the corrosive contaminants from nearby chemical plants are factors that cause the steel to corrode. Temperature also is a significant factor that increases the rate of corrosion in steel. The presence of oxygen is the most important factor in the corrosion of steel.

The writer needs to explain how the recommendations were determined, for example by citing studies to confirm these hypotheses.

A bibliography should be included.

Recommendations

One way to reduce the rate of corrosion to exposed steel is to add small amounts of chemical compounds to the surface of the steel. Inorganic anions such as polysulfates, silicates, borates, and phosphates are typically used.

The use of chemical compounds can create problems even though they may retard corrosion. If incorrect compounds are used, corrosive penetration into the steel can accelerate. The use of chemical compounds to prevent corrosion requires continued monitoring to verify that the compounds are being continuously added.

Corrosion protection can be supplied to tanks, piping, and other equipment by installing an internal solid plastic lining. The materials for lining range from natural rubbers through synthetic rubber, and include other plastics.

7

FIGURE 23–2
(*continued*)

KEY CONCEPTS

- Although the results of research may not have an immediate application, future generations depend on them to solve yet undetermined global problems.
- Our standard of living and individual expectations for comfort and achievement depend on the results of research.
- Analytical skills can be developed by using research techniques.

STUDENT ASSIGNMENT

1. Submit a formal research report to your technical writing instructor for technological or sociotechnological research completed for one of your engineering classes, or a work-related project. You may deviate from the format recommended in this chapter (shown in Figure 23–1) to meet the requirements of your engineering instructor or employer.

   Submit a clean copy (i.e., without instructor comments) of this report to your technical writing instructor. Although the accuracy of this report will not be part of your grade, your presentation of this report and your ability to technically express your results coherently and logically will form the basis for your grade.

2. Assume that our body of scientific knowledge is limited to the science and mathematics textbooks used for your introductory classes. Write a research paper explaining one of the following concepts:
   a. How catalysts operate.
   b. The stability of two-wheeled vehicles.
   c. The cause of ocean waves.
   d. How static friction is created in carpets.
   e. Why light rays refract in water.
   f. The effect of temperature on the fatigue strength of metals.
   g. The rate of evaporation of seawater.
   h. The cause of aerodynamic drag.

3. The year is 2050 A.D. Smog in your city has caused industry to leave, which has resulted in an economic recession. You are the Chief Engineer of the Save the Environment and Economy from Disaster (SEED) Foundation and have received a grant to study the long-range effects of this problem. Develop a research program to study the effects of this sociotechnical problem, and report your findings.

4. The government anticipates that by the year 2100 A.D. we will run out of hydrocarbon fuels—oil, gas, and coal—and that nuclear power will be prohibitively dangerous. Production of plastic is now a problem without hydrocarbons. Develop a research program to predict the change in our life-style.

Communication of the Self

Sections I through III give you the ability to write an effective technical report for your intended audience. Section IV expands your ability to communicate in contexts different from the previous sections.

Chapter 24, "Technical Presentations," discusses the types of presentations that technical professionals deliver, preparing the text and visual aids for these presentations, and delivery to your audience.

Chapter 25, "Career Search Communication," provides the written communication skills for finding a job: the résumé, cover letter, and thank you letter.

After completing Section IV, your ability to communicate in all technical contexts will be enhanced.

Technical Presentations

Your greatest opportunity to persuade others is through technical presentations.

After a colleague delivered a dynamic lecture, I asked him how long he spent preparing for it. His humorous response caught me off guard. He replied, "I began working on this talk in kindergarten, but it took me all these years to completely understand what the audience needed to know." His words are the key to all successful presentations: know your subject and know your audience.

This chapter cannot help you understand your subject; that is your responsibility as a professional. However, this chapter will help you make an effective oral presentation by giving you the tools to understand and meet the needs of your audience.

The ability to make effective presentations cannot be overemphasized, because it should prove to be more important than any other ability for your professional development.

TYPES OF PRESENTATIONS

The responsibility for making technical presentations arises to meet different organizational needs as described in this section (see Figure 24–1).

Informational

As a technical professional, your presentations will generally consist of the transfer of information to the audience, such as progress reports, technical papers or research, reports on the results of a test program, or descriptions of a procedure. An informational presentation must be delivered in a sequence that builds on previous information and at the level of understanding of the audience.

FIGURE 24–1
Types of Technical Presentations

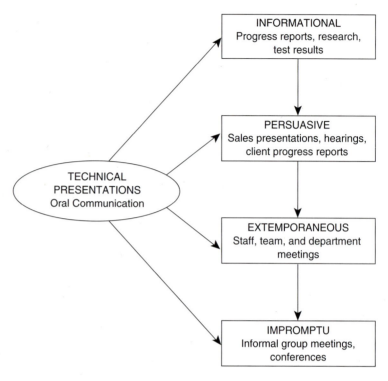

Persuasive

Sales presentations, hearings, and presentations to clients and management are intended to persuade the audience. These presentations must be delivered to the audience with energy and enthusiasm to communicate a feeling of confidence. The participants in the audience are frequently receptive to persuasion; however, at times they may be adversarial, such as at a city council meeting. In either case, the burden is on you to demonstrate the benefits of your plan or proposal.

Extemporaneous

Most organizations have regular staff meetings to determine progress on projects. In these meetings, and others, each participant is expected to discuss the events since the previous meeting. Because your participation is expected, you have adequate time to prepare your thoughts and notes. However, at any time during this informal presentation, you should be prepared to answer questions and respond to comments relating to your project. Your preparation for these presentations should include discussion of recent events, with emphasis on problems that have not yet been resolved and related issues that may arise.

Impromptu

Impromptu presentations usually occur either at informal group meetings or at conferences when additional information is needed. The key to a successful impromptu presentation is to organize your thoughts and to channel anxiety (see "Overcoming Anxiety" and Figure 24–12, later in this chapter). These presentations are probably the most difficult to deliver because little or no time is available for preparation.

PREPARING FOR YOUR PRESENTATION

An effective presentation requires thorough preparation. Your first formal presentation will consume hours of preparation for you to develop self-confidence in knowing your subject matter, to understand the needs of your audience, and to plan your presentation. Subsequent presentations of a similar type will not be as time-consuming but will also require thorough preparation. The procedure described in the following paragraphs and in Figure 24–2 can be used as a guideline for preparing your presentation.

FIGURE 24–2

Preparing for Your Presentation

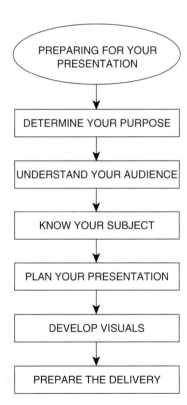

1. **Determine your purpose.** Every presentation has a purpose. Evaluating your objective is the first step in planning your approach to the presentation. Is your main objective to provide information such as the explanation of a technical concept or progress on a project, to persuade your audience such as in a sales presentation, or to be granted the approval of plans?

   An informational presentation must be delivered in a sequence that the audience can follow easily.

   A persuasive presentation needs to explain relevant facts and data that the audience may need to know to be convinced of your point.

2. **Understand the needs of your audience.** Before you begin preparing the delivery of your presentation, answer the following questions concerning your audience:

   - What is the professional background of the audience? Engineers and scientists? Technicians? Managers? Nontechnical general public?
   - What does the audience expect to learn from your presentation? Facts and data? A new procedure? An overview of a new concept?
   - How will the audience use the information? To make a decision? To further develop a concept or theory? For information only?

     The needs of the audience frequently differ from your purpose. For example, you may intend a progress report to a client to be information only; however, your client may be using this information to determine whether similar projects will be feasible in the future.

   - What method of presentation does the audience respond to? Lecture? Persuasive argument? Audience interaction with the speaker? Visual aids?

3. **Know your subject.** It is important for you to be knowledgeable in the subject of your presentation so that you understand what the audience needs to know. Because presentations usually are the result of an intensive project or study, you will probably be familiar with the information. However, you should not rely on this familiarity. All relevant subject material must be carefully reviewed before you begin to prepare the outline of your presentation.

   To establish your credibility as an informed speaker, you must communicate your thoughts and ideas in a self-confident manner. When you are not thoroughly familiar with the subject of your presentation, this will be revealed to the audience in your voice, gestures, and the content in the first few moments of your presentation, and you will lose the attention of the audience.

   After you have learned your subject matter thoroughly, try to anticipate the questions that the audience might ask and prepare to answer them. Keep in mind that, occasionally, out of curiosity, a question might be asked that is not directly related to your presentation. When that occurs, courteously respond that the answer to that question is beyond the scope of your presentation and that you would be happy to discuss it in more detail after you conclude your presentation.

4. Plan your presentation. Do the following to plan the delivery of your presentation:

a. Organize your information.

    i. Review all relevant subject material, and determine the main point from which you will develop your presentation.

    ii. Gather your supporting information, and determine three or four subpoints that you will use to validate your main point. Be selective. An audience can remember no more than three or four ideas or concepts and will remember only those that it considers to be relevant.

    iii. Consider the time allotted to make your presentation. The ability of the members of the audience to comprehend your message diminishes considerably when your scheduled time has elapsed. Also, it is inconsiderate to subsequent speakers.

    iv. Select the structure for your presentation:

Use a chronological approach for procedures and instructions, such as operating machinery.

Use cause-and-effect for explanations, such as the theory of fatigue failures.

Use a general-to-specific approach for descriptions, such as a sub-division layout.

Use a list for parallel, but unrelated, items, such as recommendations for reducing the cost of a product. The list should be in descending order of importance.

Use comparisons for equivalent, but different, items, such as alternative proposals for a rail transportation system.

Use advantages and disadvantages for persuasion, such as convincing management to purchase state-of-the-art surveying equipment.

Many presenters use a combination of the preceding structures to communicate the different components of their messages.

    v. Summarize the main point and each subpoint of your presentation into a word or short phrase. Use the structure selected from the preceding list and develop an outline for the body of your presentation.

b. Develop the presentation.

    i. Develop the text of the body of your presentation from the outline you developed. Some speakers like word-for-word text, while others feel comfortable with only the outline. To emphasize the important points of your presentation, prepare to repeat each point from different perspectives during your presentation.

    ii. Prepare an introduction to the body of your presentation. Tell the audience what you will be speaking about and how they can use this information. Include any background material that the audience will need to understand your presentation.

iii. Prepare a closing summary. Review your main point, and include any action or recommendation that you would like your audience to follow up.

iv. Determine which visual aids you will use to help the audience visualize and remember your main points and to keep their attention. You may select overheads, slides, videos or movies, flip charts, chalkboards, handouts, and computer-generated projections.

v. Prepare an outline of key words and phrases for your presentation on 3×5 index cards.

5. **Develop visual aids, and select your visual tools.** Visual aids in formal presentations keep the audience's attention, reinforce your message, and help the audience visualize and remember your main points. These guidelines should be followed for developing visual aids:

- Use large and clear visual aids that the audience can see from anywhere in the room. Use a large type size for headings.

- Use color for greater impact.

- Use pictures, charts, graphs, and sketches rather than words and numbers, because they are easier for the audience to interpret.

- To have the greatest impact, demonstrate only one idea or concept with each visual.

- Keep each visual simple. Too much information or clutter will detract the audience's attention.

- Be accurate, and be consistent in your visual aids with the information presented in your talk.

- Use visual aids as a supplement, not the basis, of your presentation. Do not let the visual aids dominate your presentation or direct the attention of the audience away from you. Use visual aids only for emphasis.

- Use visual aids that are appropriate for the technical level and interests of the audience.

The advantages and disadvantages of the most commonly used visual tools for displaying your visual aids and the most effective methods for using these tools are discussed in the following list and in Figure 24–3.

- Overheads (see Figure 24–4). Overheads allow the speaker to have complete control of the audience because its attention is directed to a screen adjacent to the speaker in the front of a well-lit room. You can control the introduction of information by adding overlays to the original transparency, by blanking out and exposing portions of the trans-

FIGURE 24–3
Visual Tools

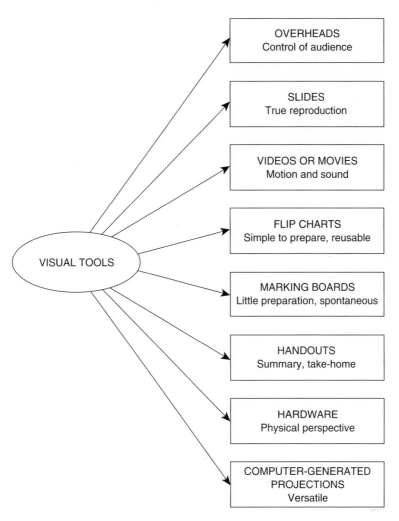

parency as desired, and by writing on the transparency with a felt-tipped marker. When using a pointer, use the projection on the screen, rather than the overhead projector, as your work area to avoid blocking someone's view.

- Slides (see Figure 24–5). Slides are an excellent visual aid when true reproduction is desired. Slides allow you to organize your material and to control its presentation from the front of the room. However, a slide presentation requires a dark room, which breaks the rapport between the audience and the speaker. This disadvantage can be overcome by speaking from a lighted lectern that is visible to the audience, by showing no more than five or six slides in sequence, and by introducing each slide before it appears on the screen and telling the audience what to look for.

FIGURE 24–4
Overheads

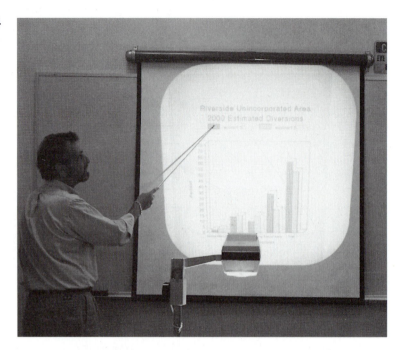

- Videos or movies (see Figure 24–6). Videos and movies demonstrate motion and sound to the audience. Similar to slides, they are a true reproduction. However, videos and movies essentially replace the speaker and break the rapport with the audience. This problem can be avoided by showing your video or movie in short segments and introducing each segment to the audience before its presentation. When you use a video or movie, it should be a small part of the total presentation.

- Flip charts (see Figure 24–7). Flip charts are simple to prepare, are easy to use in a well-lit room, can be modified as you speak, and allow you to return to information previously discussed. The information on the charts should include basic points only and should be easy to read. To prevent losing the audience's attention between references to the flip chart, you can place a blank sheet of paper between visuals, and mask each visual when you finish discussing it.

- Marking boards (or chalkboards) (see Figure 24–8). Marking boards (or chalkboards) require little preparation, facilitate the explanation of spontaneous remarks, and are inexpensive. However, the marking board has many disadvantages for an effective presentation. It is time-consuming to use, the speaker's back faces the audience when information is written on it, the writing is sometimes illegible and disorganized, and the information cannot be removed easily. These disadvantages, except for removing the information, can be overcome by carefully

FIGURE 24–5
Slides

FIGURE 24–6
Videos or Movies

FIGURE 24–7
Flip Charts

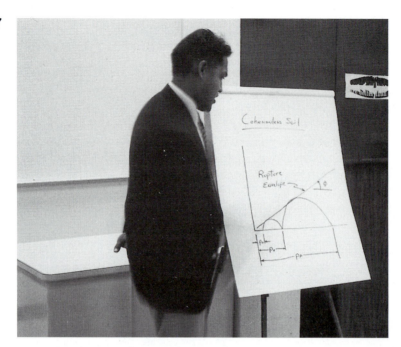

FIGURE 24–8
Marking Boards or
Chalkboards

planning the information that will be be placed on it or by writing the information on the marking board before the presentation.

- Handouts (see Figure 24–9). Handouts are a convenient medium for an outline of your presentation, a summary of intended follow-up activity for the audience, and supplementary material and data to support your message. An outline should be distributed to the audience at the beginning of your presentation. Other handouts should be distributed at the conclusion of your presentation to prevent the audience from reading the handouts during your presentation.

- Hardware/prototypes (see Figure 24–10). Physical hardware, such as a printed circuit board or sample of material, and prototypes (scale models), such as a model of a house, are excellent visuals to give the audience the perspective of size, weight, and feel—a perspective that is not available with any other visual. The hardware or prototype should be obscured from view when it is not being discussed and made available for inspection at the conclusion of the presentation.

- Computer-generated projections (see Figure 24–11). With the appropriate equipment and presentation software, the image of any computer display can be projected onto a screen in front of the room or onto individual video displays at each station. These images can be controlled from anywhere in the room with a remote device. These computer displays can include spreadsheets, scanned photographs, computer-generated slides, and computer animation with sound.

Computer technology is changing so rapidly that it is difficult to predict innovations in the use of computers for your presentations. However, the effectiveness of your presentations can be dramatically enhanced when you take advantage of the most advanced computer capabilities.

6. **Prepare the delivery of your presentation.** After you have developed your materials for your presentation, do the following to prepare for your delivery:

   a. **Check the Environment**

   Visit the location where you will be giving your presentation.

   i. Verify that the lighting is adequate, the size of the room and the number of available seats are appropriate for the anticipated size of the audience, electrical outlets are conveniently located, table space for your materials is available, and the lectern is comfortable.

   ii. Choose a seating arrangement so that latecomers enter the room at its rear and create a minimal disturbance. Also, determine whether any competing noisy activities will occur in adjacent areas. If so, consider alternatives such as noise barriers for reducing disturbances. When you judge these alternatives to be inadequate, select a different room for your presentation.

FIGURE 24–9
Handouts

FIGURE 24–10
Hardware/Prototypes

FIGURE 24–11
Computer-Generated Projections

iii. Slowly pace the "stage" area back and forth to develop a feeling of confidence.

iv. Check the microphone (if needed) and the visual equipment you will be using for your presentation to make sure that you know how to operate them and that they are working properly.

b. Rehearse

i. Review and revise, if necessary, the key words and phrases on the 3×5 index cards prepared previously.

ii. Mentally rehearse your presentation using the index cards until you are ready for a sound rehearsal.

iii. Rehearse your presentation several times using the equipment and visuals that you will use for your live presentation.

iv. Deliver one or two dress rehearsals in front of several friends, an audiotape, or a videotape. Revise your presentation, if necessary, to make it flow smoothly.

v. Anticipate potential questions, and prepare to answer them.

DELIVERING YOUR PRESENTATION

You have spent hours preparing for your presentation and are now ready to deliver it. As you begin your presentation, do not make impromptu apologies such as, "Bear with me while I get my notes organized" or "I'm sorry that I'm late, but traffic was awful this morning." These remarks imply a lack of preparation and create an immediate barrier between you and the audience. The suggestions discussed in this section will help make your presentation successful.

Overcoming Anxiety

For an inexperienced speaker, anxiety is always the foremost concern. Anxiety can be channeled by doing the following (see Figure 24–12):

Before the Presentation

- Reassure yourself before the presentation that you have something of value to offer the audience, they are interested in listening to you, and you have adequately prepared for a successful presentation.
- Visualize yourself self-confidently walking into the room and being introduced, delivering you presentation with enthusiasm, competently answering questions, and leaving the room knowing you did a great job.

During the Presentation

- Concentrate on speaking slowly.
- Stand erect and breathe deeply. Pause and take a deep breath whenever you begin to feel anxious.
- Release tension by occasionally using hand and arm gestures and taking a short step to the side while you are speaking.

Body Language

The movements of your body can either encourage or lose the audience's attention. The following suggestions will help you keep its attention (Figure 24–13):

- Stand erect but try to be at ease. Do not distract the audience by shifting your weight from side to side.
- Occasionally move one or two small steps to the side to help keep the attention of the audience, but do not pace back and forth.
- Take one or two steps toward the audience to help you stress an important point. However, to avoid intimidating the audience, always keep at least six to eight feet between you and them.
- Use natural hand and arm gestures for emphasis. However, avoid fidgeting with your lecture materials, or keeping your hands in your pockets, behind your back, or on your hips. Do not cross your arms in front of you.
- Wear suitable business clothes for any formal presentation.

Voice

The common problems of voice delivery are usually caused by anxiety. These problems can usually be resolved by audiotaping, evaluating, and rehearsing your presen-

FIGURE 24–12
Overcoming Anxiety

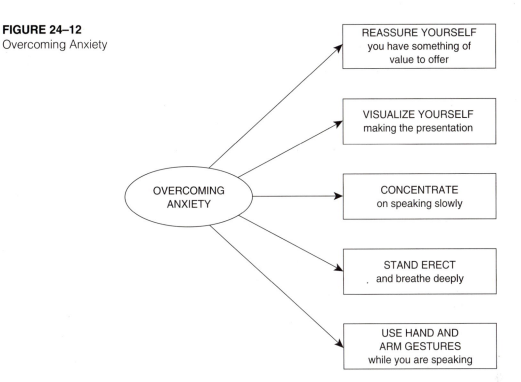

tation until you feel that your delivery demonstrates self-confidence. Your voice can project this sense of self-confidence when you do the following (see Figure 24–14):

- Speak slowly, enunciate your words, and pause occasionally to allow the audience to grasp the meaning of your message.
- Vary the tone, volume, and rate of speed of your voice to maintain audience interest. Avoid a monotone voice.
- Project confidence by understanding the importance of your message to your audience.
- Make sure that your volume is appropriate for the size of the audience and room. Your voice should be easily heard in the back of the room without drowning out the audience in the front. Avoid trailing off the volume of your voice at the end of sentences.
- Refer to your notes only to refresh your memory of key words and phrases. Unless you are quoting an authority, avoid the temptation of reading your notes.
- Deliver your subject matter with enthusiasm by smiling and using facial gestures.

Audience Interaction

As you speak, move your eyes to various parts of the room, and when the audience is small enough, make eye contact with each person for several seconds. This will establish rapport with the audience and get them involved in your pre-

FIGURE 24–13
Body Language

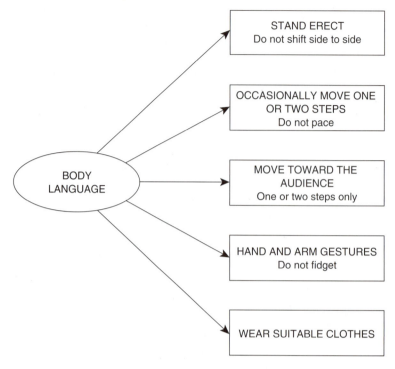

sentation. It also gives you the opportunity to interpret the audience's response to your presentation and allows you to react accordingly. Also, if it feels comfortable for you, include humor in your presentation.

Answering Questions

When you prepare your presentation, you should anticipate that the audience will have questions when you finish speaking. The audience should be encouraged to ask these questions so that they leave the presentation with a confident feeling (see Figure 24–15).

- To avoid giving the impression that you have completed your presentation and are ready to leave, do not put away your visuals or other materials before you answer the questions from the audience.

- Encourage questions by asking, "What questions do you have?" Raise your hand as you say this to give the audience a cue for doing the same.

- Listen carefully to each question to make sure that you understand its intent. Always allow the person asking the question to finish asking it.

- Repeat each question before you respond to it. Do this to make sure that you understand it and also because, more than likely, some people in the audience may not have heard it.

FIGURE 24–14
Voice

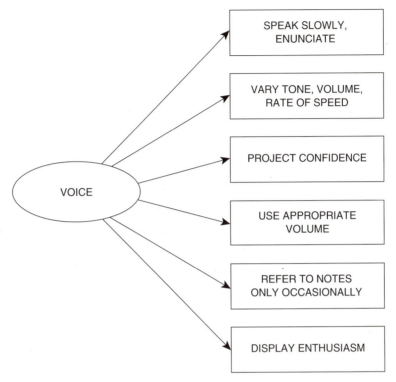

- Do not get upset when you do not know the answer to a question. No one person can know everything. When this happens, confidently admit to your lack of knowledge, and tell the asking person that you will research and communicate the answer in the near future.

- Even though only one person has asked the question, focus your attention on the entire audience, because others may have the same question, and you do not want to lose your rapport with the audience.

SPECIAL APPROACHES FOR EXTEMPORANEOUS AND IMPROMPTU TALKS

As a professional, you will be asked frequently to make extemporaneous talks (prepared in advance, but delivered with few or no notes) and impromptu talks (presentations with little or no preparation) at meetings and conferences. This is not a reason to panic when you understand the appropriate techniques for handling the situation.

Except for the additional time available for preparing an extemporaneous talk, the preparation and delivery of these talks are similar because neither requires the use of visual aids, and you have no control over the presentation environment.

FIGURE 24–15
Answering Questions

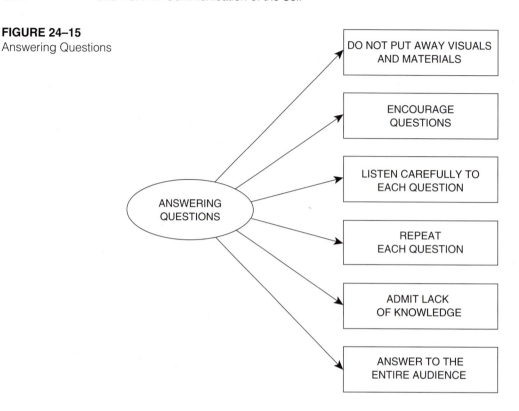

Follow these suggestions for an extemporaneous or impromptu talk (see Figure 24–16):

1. **Prepare**
   a. Determine whether your talk should emphasize one function, such as design, fabrication, or management, or whether several functions should be addressed. Then, for an extemporaneous talk, gather your information. For an impromptu talk, you will rely primarily on what you know because preparation time is not available.
   b. Decide on a structure for your approach: chronological, cause-and-effect, advantages and disadvantages, etc.
   c. Select the main points that you want to discuss and write them down.

2. **Talk**
   a. Take your position in the room for addressing the group.
      i. For small meetings and conferences, you will ordinarily have the choice of remaining seated for your presentation, standing at your seat, or coming to the front of the room. Many speakers feel most comfortable with the informal atmosphere created when they remain seated, and others prefer the control of audience attention when standing or coming to the front of the room. Sometimes, a

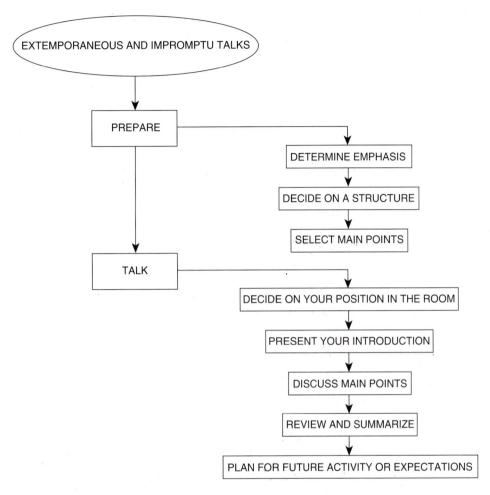

FIGURE 24–16
Extemporaneous and Impromptu Talks

           visual aid such as a marking board will determine whether you come to the front of the room.

ii. For large meetings and conferences, you should always either stand at your seat so that you can be heard throughout the room, or come to the front of the room when your presentation is lengthy.

iii. When you are not the first speaker at these large meetings and conferences, to keep the attention of the audience on your topic rather than distract them by altering an established routine, follow the positioning example of those who speak before you unless you have a reason to do otherwise.

b. Introduce your talk with a comment concerning your background with respect to the topic. When additional information would be helpful to the audience concerning the appropriateness of your talk at this time, include this in your introduction.

c. Discuss the main points of your talk. When your topic is controversial, acknowledge other points of view before you begin your discussion.

d. Review the main points, and summarize what you have told them.

e. If applicable, conclude your talk with a plan for future activity or expectations.

GROUP CONFERENCES AND MEETINGS

Group conferences and meetings are becoming a commonly used vehicle for solving interdisciplinary technical problems. Undoubtedly, they are time-consuming, draw unclear lines of authority, and are sometimes politically counterproductive. However, when group conferences and meetings are properly run, the expertise of several professionals can be blended to have a synergistic effect. Another advantage of group conferences and meetings is that participants are usually more willing to become team players to complete the project.

At a group conference or meeting, an atmosphere of harmony is created when all members have equal status and are encouraged to offer ideas and opinions, even when they are controversial. This harmony can be achieved when all participants do the following (see Figure 24–17):

- Understand the purpose of the meeting and its potential value.
- Listen carefully to the contributions of the other participants.
- Be considerate of the ideas and opinions of others. Assume that these ideas and opinions are offered in good faith and without ulterior motives.
- Do not interrupt the other participants when they are expressing their points of view.
- Support any ideas or opinions that have any merit. Recognize that the real benefit of brainstorming usually occurs when novel, but impractical, ideas are modified to make them workable.
- Feel free to disagree with the ideas and opinions of the other participants, but never attack them personally.
- Offer suggestions for improving another's idea rather than only explaining the reason it is unworkable.
- Do not become defensive when the lines of communication break down and individuals are ridiculed for one of their ideas or opinions.
- Encourage others to express their ideas and opinions, and discourage criticism of others.
- Do not dominate the other participants.

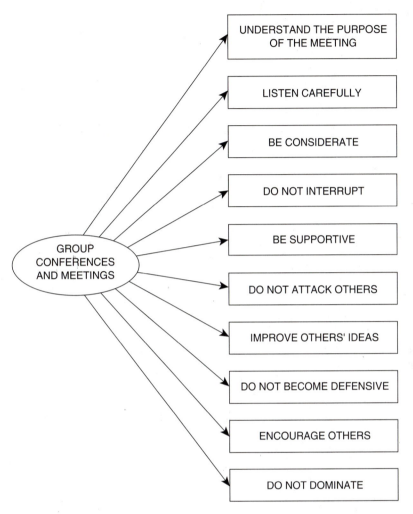

FIGURE 24–17
Group Conferences and Meetings

KEY CONCEPTS

- The ability to make effective oral presentations is undoubtedly the most important skill for the career development of a professional.

- It is only a myth that great speakers are born, not made. The ability to speak effectively in front of an audience comes with developing and practicing the skills in this chapter.

- Anxiety is a natural emotion for all speakers, especially inexperienced ones. The ability to speak comfortably in front of an audience can be achieved by reassuring yourself that you have prepared sufficiently to make a presentation that will generate interest among the audience and that they will benefit from.

STUDENT ASSIGNMENT

1. *Informational presentation.* Deliver a 5-minute presentation on a report written for this class selected from one of the following:

   - An activity or product evaluation to the management of your organization
   - A feasibility report to a prospective actor or purchaser
   - An environmental impact report to a city council
   - A technical article to an audience comprising professionals in all technical disciplines
   - A laboratory report, analysis, or research report to your colleagues

2. *Informational presentation.* Deliver a 5-minute (or 10-minute) presentation to a technical audience on a technical paper or innovative project. You may use a section of any technical textbook or an article from a technical magazine or journal as the basis for your presentation.

3. *Persuasive presentation.* Deliver a 5-minute (or 10-minute) presentation to your city council or board of trustees of your college or university (as applicable) to persuade them to approve a proposed course of action. You may propose one of the following:

   a. A solution to any local sociotechnical issue such as traffic congestion, recycling, or overcrowded dorms (use any current newspaper, magazine, or available data from the police department, etc. as your source of information).

   b. A course of action that would benefit you or your company such as permitting a deviation from a zoning ordinance to allow a multiple-dwelling or temporarily closing off a thoroughfare to expedite construction.

   You may use handouts or other visuals.

4. *Sales (persuasive) presentation.* Your company has been invited by a prospective client to make a sales presentation to demonstrate the advantages of owning or using the product, project, or service that your company offers. An associate and you have the responsibility of delivering a 20-minute sales presentation to technical, managerial, and administrative personnel of the prospective client. The presentation will be followed by a short question and answer period. Visual aids are to be used for the presentation. Technical information can be obtained from manufacturers, government agencies, sales brochures, periodicals, etc. Your attire is to be appropriate for a formal sales presentation.

The presentation should emphasize the technical, sociological, and economic (increased efficiency) advantages, and should not address price or competition. Typical choices might include a high-performance vehicle, mass transit, computer software, or design services.

5. *Extemporaneous presentation.* Prepare yourself at home to make a 10-minute extemporaneous presentation to a peer group on a technical procedure (e.g., a laboratory or analytical procedure) or project (e.g., a class or work project) that you are familiar with. Lecture notes are to be minimal, and the marking board or chalkboard is to be your only visual aid.

6. *Impromptu presentation.* With no more than 5 minutes for preparation, make a 2-minute impromptu presentation on one of the following topics selected by your instructor:

The term *organization* in a through f includes any place of employment or volunteer service, present or past.

    a. Explain a technical or managerial procedure at your organization.

    b. Describe a problem at your organization and how it was remedied.

    c. Explain the advantages and disadvantages of working for your organization.

    d. Explain how the competition would describe your organization.

    e. Describe your favorite work-related activity at your organization, and explain why.

    f. Explain what you would change at your organization to make its operation more efficient.

    g. Explain which invention has had the most impact in your field of study and why.

    h. Explain a process or invention that has enhanced the quality of your life and why.

    i. Discuss your favorite high-tech movie.

    j. Discuss your favorite high-tech personal possession.

    k. Describe your dream home.

    l. Describe your dream car.

    m. Describe a process or invention to solve a sociological problem.

    n. Describe a high-tech device, not yet invented, to deter crime.

In Student Assignments o through t, predict and discuss, as directed by your instructor, the following for (a) 5 years from now, (b) 25 years from now, or (c) the 22nd century:

    o. The work station.

    p. The process of higher education.

    q. The stock of bookstores.

    r. The freeways and turnpikes.

    s. Home entertainment.

    t. Parking lots and parking structures.

CHAPTER 25

Career Search Communication

Transitions between professional experiences are facilitated by effective communications.

As a professional, it is likely that you will have several employment and professional experiences. The transitions between these experiences are accomplished through networking (connections with people of like experience) or through written communication and interviews by interested organizations. This chapter discusses these written communications: résumés, cover letters, and thank-you letters.

RÉSUMÉS

Résumés are not reports, but, similar to many types of reports, they are written communications submitted to professionals in other organizations for their review. Each time you change employment, you will usually be asked by prospective employers to submit your résumé as a condition for being interviewed. A well-written résumé may give you your only opportunity to present yourself to the prospective employer.

As a professional manager, you will be responsible for hiring employees who are competent and compatible with others in your organization. The most effective mechanism for screening many applicants is to review their résumés for selecting those to be interviewed. Understanding how to read résumés may help you select the most qualified applicants for these interviews.

Also, many organizations submit résumés of key employees with bid packages to potential clients to validate their performance qualifications. These résumés typically follow standard organization formats and are edited for uniformity and clarity by these organizations' personnel departments. Therefore, these résumés are not specifically addressed in this chapter.

Curricula vitae (plural of curriculum vita), commonly abbreviated as cv, is a term used in academia and some organizations for personal communications that emphasize academic and professional achievements of potential employees. Usually, this term is used when the positions require demanding credentials. These personal communications may be identical to résumés except that they include more details and are frequently several pages long.

Purpose

A résumé publicizes your availability for employment. It lists your qualifications in a concise, well-organized manner that will hopefully impress the reader.

Your résumé is used to achieve two objectives:

1. Obtain an interview with a prospective employer.
2. Serve as a basis for discussion during the interview with the prospective employer.

The prospective employer will rarely spend more than 30 seconds reviewing a résumé and will frequently reject a résumé in less than 10 seconds. Therefore, to ask to be interviewed, the résumé must be professional-looking and must readily communicate important information. A neat and organized appearance, quality printing and stationery, no typographical errors, and headings with appropriate information serve this purpose.

For the résumé to serve as a basis for discussion during the interview, the interviewer must be able to easily retrieve the pertinent information during the interview.

Preparation to Write

An effective résumé requires effort to write. You will probably need several drafts before you determine the content and format that works best for you. Begin to prepare your résumé at least 3 or 4 weeks before you need it.

Think about your career goal for the next 5 years. Evaluate the background and experience that you will need to reach that goal (the career center at your college or university can assist you). With this goal in mind, make a list of several current career possibilities (e.g., computer analyst, research chemist, instrumentation designer, construction manager, machine designer) that can help you achieve that goal.

Make a list of your accomplishments and activities including the following:

- Education and degrees
- Applicable classes completed
- Academic grade point average
- Senior project, if any
- Work experience
- Honors, awards, and scholarships

- Technical skills and licenses
- Publications
- Leadership and extracurricular activities
- Community service
- Interests and hobbies

Assemble your list into the headings described later in this chapter, and within each heading, arrange the strengths of your accomplishments and activities in descending order. Keep your immediate goals in mind, and remember that the more recent an accomplishment or activity is, the more impressive it is to the prospective employer.

Personal information (e.g., age, height, weight, marital status and dependents, health, and ethnic or religious background) should not be included on a résumé. A prospective employer is allowed by law to ask only those questions that pertain to the job requirements, and even though you may be proud of yourself, personal information may prejudice the interviewer against you. The career center at your college or university can help you determine which personal information the prospective employer can ask. If you apply for a job that requires a security clearance, you may include your citizenship. You may also include your willingness to relocate or travel.

Structure

Because the structure of the résumé is the framework for the contents and is seen by the prospective employer before the résumé is read, the structure should be attractive and easy to read.

Left, right, top, and bottom margins should be at least 1 inch wide. Emphasize headings by using bold, large, or capital letters. They are usually left justified but are sometimes center justified. The information under each heading can be directly below the heading and offset from 1/2 to 1 inch to the right (see the sample résumés in Figures 25–2 and 25–3 later in this chapter), directly opposite (not below) the heading and offset adequately so that the information opposite all the headings has the same left justification (see Figure 25–4), or below the heading (see Figure 25–5).

As a technical professional, you want your résumé to appear traditional and without artistic flair. Use only one style of font for the entire résumé. Except for the headings and your identifying information, do not use a different-size font or bold print to emphasize information. Underlining is acceptable for headings and proper nouns (e.g., the name of your college or university or the title of jobs) but should not be overdone. Your résumé should not appear cluttered.

Limit your résumé to two pages and, if possible, to one page. A one-page résumé is recommended for a new graduate with limited professional experience.

Make sure that your résumé is grammatically correct, has no spelling or punctuation errors, and has up-to-date and accurate information.

A résumé usually follows one of the following formats:

- **Chronological:** This type of résumé relates *what, where, and when,* with the most recent experience listed first. This format is the most factual and the easiest to understand, but it emphasizes gaps in employment history and inappropriate or insufficient experience. This format is the preferred format for persons with limited experience and is usually used by new graduates.

- **Functional:** This type of résumé categorizes skill areas and strengths acquired from a variety of experiences. Unless this type of résumé is carefully prepared, it is often confusing and vague to the reader. This format is usually used by persons with many years of experience who are considering a career change. However, this type of résumé is sometimes used by new graduates with varying experiences from part-time and summer work, campus organizations such as American Society of Mechanical Engineers or Eta Kappa Nu, and volunteer organizations such as the American Cancer Society.

The structure and headings of the functional résumé are the same as the chronological résumé. Generally, only the content in the experience section differs, although, in the functional résumé, content from other sections such as extracurricular activities are sometimes included in the experience section. See the critique of résumé, Sample 2 on page 400 for an example of an "EXPERIENCE" section in a functional résumé.

Content

The sections of your résumé are discussed next and are shown in the usual order of presentation (see Figure 25–1).

Heading   Your name, address, and telephone number are placed at the top of the résumé so that the prospective employer can easily retrieve them. This information can be left, right, or center justified, or your name can be left justified, and your address and phone number can be right justified. This information can be in bold, large, or capital letters. For easy recognition, your name should be more prominently displayed than the other information in the heading. Do not place the label "Résumé" on the top of the page.

As a student, you might want to include your permanent address and telephone number as well as your school address and telephone number. Make sure you label each one and that you include your area and zip codes. When you are not available at your telephone during normal working hours and you have an answering machine, indicate this immediately after your telephone number with "(message)." However, remember that your outgoing message must be serious to reassure the caller, your prospective employer, of your professional demeanor.

Summary   Employers, because of the many résumés that are received in response to an employment opening, appreciate a summary at the beginning of a résumé to determine the applicability for the opening of an applicant's background and experience. This summary eliminates the need for the reader to inter-

FIGURE 25–1
Sections of a Résumé

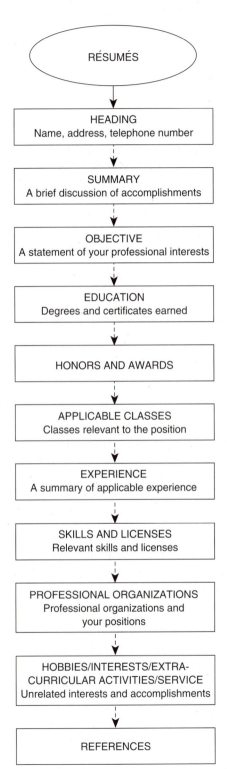

pret the contents of the résumé and therefore saves this reader time and, more importantly, prevents the reader from misinterpreting or overlooking important qualifications of the applicant. You may need more than one résumé to emphasize the appropriate background and experience for different types of openings.

The summary should include applicable items of accomplishment and should not include explanations, which more appropriately belong in a cover letter, or objectives. It can be written in either sentence or phrase format. It should be no longer than several lines.

A sample summary in sentence format follows:

I have 6 years of recent avionics testing experience, 4 years of avionics design experience of commercial aircraft, and 3 years of project management experience. Also, I earned a Master of Science in Electrical Engineering from Purdue University.

A sample summary in phrase format follows:

Design of automobile brake components, 3 years; test of automobile suspension systems, 2 years; knowledge of CAD. BS in Mechanical Engineering, 3.4/4.0 GPA.

Objective   Many résumés include an objective to tell the prospective employer your interests. The objective should be clear, concise, and specific. It can include the professional field or industry, job title, or desired skills. When you are interested in part-time or summer employment only, mention this in the objective.

When you send résumés to many organizations, you will probably need several different résumés. This is because even though your background may be appropriate to meet the needs of the organization, an objective at the beginning of the résumé that is incompatible with the organization's needs may deter the reader from getting to the relevant information.

For example, you desire to work in analysis and express that desire in your objective; however, you have experience in testing and are willing to work in that discipline again. Unknown to you, the organization has an opening in testing, but the reader fails to read your experience section.

For these reasons, the objective is frequently omitted from the résumé or is included in a cover letter only after a personal discussion (usually by telephone) with the prospective employer.

Example objectives follow:

An entry-level position as a stress analyst in the aerospace industry.

Summer employment in the communications industry.

A design position with a multidiscipline civil engineering company.

Employment in research chemistry in a pharmaceutical company.

The following sample is vague and should be avoided:

A mechanical engineering position with a potential for growth into management.

The section immediately following your objective (if included) should present your most positive asset for obtaining the desired employment. For the new graduate, this is usually education. In the future, after your first professional experience, experience will precede education.

Education   For each academic experience, include your degree, college or university, city and state, date of graduation, major, minor or option, and grade point average (when B or above). When you have attended more than one college or university, list your most important academic experience first (e.g., a master of science earned several years ago is usually more important than, and should be listed before, a recently earned certificate or associate of arts degree. Unless you are a freshman or a sophomore, do not include your high school education.

The title of your degree (e.g., Master of Science in Electrical Engineering) should precede the name of the school for all academic experiences unless

- You want to emphasize that you have attended a high-profile school such as Harvard or Stanford.
- You do not want to emphasize that you did not receive a degree from every institution indicated.

In either of these cases, the name of the school will precede your degree (if any) for all academic experiences.

If you are a student applying for a job contingent on your future graduation, the date of graduation is presented as if it were a past event; that is, you do not need to add "(anticipated)" after the date, because this is self-evident by the future date shown.

When you include your grade point average, in fractional format, you should indicate this average with the basis on which it is calculated (not all schools use a 4.0-point system) and indicate whether it is your cumulative or major grade point average. For example, "3.2/4.0 GPA (cumulative)."

Honors and Awards   Any honors (e.g., Dean's List), honor societies (e.g., Tau Beta Pi), and awards are listed in this section.

Applicable Classes   Any classes that are applicable to your career objective can be listed by using a descriptive title (e.g., Active Filter Design) rather than a class number. Applicable classes are useful when you are looking for a job before graduation or you have minimal practical experience and you want to emphasize the strength of your education. You may also indicate the topic of your senior project when you completed one.

However, this section is not usually included when your work experience is applicable to the position you are applying for and your degree is in a traditional curriculum (e.g., chemical engineering or physics).

Experience   Because new graduates generally use chronological résumés, the discussion in this section pertains to that format.

Your prospective employer is interested in all of your work experience (except jobs in high school) for the following reasons:

- Your work experience may be applicable to the job you are applying for.
- Your work experience may demonstrate leadership potential or stability.

- Work experience demonstrates responsibility for financing a portion or all of your education.
- Extensive work experience may be justification for mediocre grades or a delayed graduation date.

Experience usually includes your most recent job listed first. For each job, the following are listed: your job title; company name, city, and state; dates of employment; and your job description. You can include part-time and temporary jobs, internships, and military experience in this section, but you should include volunteer work in the community service section of your résumé unless the skills obtained from this work can be used for the prospective employment.

Follow the rules below for writing a job description:

1. Prioritize your job responsibilities and duties.

2. List related responsibilities and duties together, but separate them with commas and semicolons. Unrelated responsibilities and duties are separated with periods. The responsibilities and duties most relevant to your job objective should be listed first. Sentence fragments and phrases are acceptable.

3. Use definitive action verbs such as *designed* or *researched* in describing your responsibilities and duties. Avoid passive and vague phrases such as *responsible for* or *duties included*. Also avoid personal pronouns such as *I*.

4. Describe your responsibilities and duties from a technical rather than a social perspective. Use "designed a 2-story steel building" instead of "designed an apartment building."

5. Be quantitative to emphasize your accomplishments. Use "reduced cost by 6.5%" instead of "developed cost reduction program."

6. Use generic descriptions instead of proper nouns, acronyms, or names. Use "ran performance test program for 1-ton soft-terrain vehicle" instead of "ran performance test program for Sandpiper-2."

7. Use the past tense for all activities unless they are still in progress.

8. List several job experiences that are not relevant to your job objective together as "Miscellaneous" or "Other Experience," for example, "Department store inventory clerk; pizza deliverer."

Skills and Licenses   Any special skills such as computer literacy, licenses such as certification as a welder, or knowledge of a foreign language should be listed here. However, you should include only those skills, licenses, and foreign languages that relate to your job objective (e.g., becoming certified as a calligraphist may require considerable skill and preparation, but this would typically be unrelated to your job objective and may be included in the hobbies or interests section of your résumé.

Professional Organizations   As a student, you may be a member of one, or more, student chapters of professional organizations such as the American Institute of Chemical Engineers. Professional organizations, offices held, and dates are included in your résumé. Membership demonstrates to the prospective employer an interest in your profession. Holding an office in your organization demonstrates a potential for leadership.

Hobbies/Interests/Extracurricular Activities/Community Service   These categories, as applicable, give the prospective employer a more complete picture of you. Although the listed items rarely form the basis for offering you a job, they are sometimes used to open the conversation at the beginning of the interview and can therefore be very beneficial to create a personal relationship. Your presentation of these items should be brief.

References   Because companies have recently become legally liable for giving unfavorable recommendations to former employees, many prospective employers do not rely on references. However, a prospective employer may ask for work, professional, or personal references. Before giving the names of references when applying for a position, the permission of these references must be obtained.

Many job applicants include "References Available Upon Request" on their résumé. This statement does not include any information of value to the prospective employer and merely wastes space on your résumé.

Proofreading

After you have written your résumé, one or more persons should proofread it for you, preferably, professionals employed in the same field as your prospective employers (your career center at your college or university will also proofread your résumé). If the proofreaders have difficulty understanding a portion of your résumé, the prospective employer likely will have the same difficulty. In that case, you should revise the résumé and ask the proofreader to review it again.

Getting Copies

Your résumé should either be professionally printed or printed on a laser printer. An off-white, quality bond stationery of medium-heavy weight is preferred by most job applicants. You will probably need a minimum of 50 copies of your résumé.

Critique of Sample Résumés

The headings in all the samples (see Figures 25–2 to 25–5) include the name, address, and phone number at the top of the résumé. Each résumé prominently displays the name of the applicant by using bold and large print, or separating it from the other information. Capital letters are used effectively in all samples. The graphic detail of the horizontal line in the headings of Samples 1 and 2 (see Figures 25–2 and 25–3) may help attract the attention of the reviewer. All samples are either professionally printed or, as is more common, printed on a laser printer to facilitate customizing the résumé for the various intended recipients.

Critique of Résumé—Sample 1

Sample 1, in Figure 25–2, is neat and attractive. The headings are underlined and left justified. The information is offset to the right. Allowing at least 1/2 space between the headings and the information prevents a cluttered look.

Each of the details of the work experiences should begin with an action verb rather than "Duties include." For example, the first experience could read "Draft and design subdivision infrastructure." Notice that the present employment is appropriately expressed in the present tense.

Critique of Résumé—Sample 2

The headings in Sample 2, shown in Figure 25–3, are left justified, and the information is offset 1 inch to the right. This résumé appears somewhat crowded because of the small size of the font. Perhaps this résumé should have used a larger font size and continued on a second page.

The graduation date does not need to be prefaced by the word *graduating*.

The work experience appropriately includes details of responsibilities and dates. Similar to Sample 1, each of the details of the work experiences should begin with an action verb rather than "Duties included." Otherwise, this résumé is generally well-written.

Although this student is a new graduate, she has had a variety of experiences and could have written a functional résumé. The experience section of this functional résumé would be different and might read as follows:

EXPERIENCE

Management

- Coordinated project with owner.
- Advised on preliminary design for estimating.
- Organized construction materials for form setters.
- Translated instructions to Spanish-speaking laborers.
- Cashiered 12 automated gasoline pumps.
- Inventoried automotive parts.

Construction Technology

- Designed grading plans.
- Designed concrete thoroughfare.
- Interpreted concrete drawings for form setters.

Construction Labor

- Operated heavy equipment.
- Inspected pavement settlement.

DOUGLAS J. DENNINGTON

7880 Whitegate Avenue
Riverside, CA 92506
(714) 780-0444

EDUCATION

Bachelor of Science degree in Civil Engineering; California State Polytechnic University, Pomona; June 1995

EXPERIENCE

Action verbs should replace passive phrases such as, "Duties include the drafting." More effective and concise is, "Drafted and designed."

ENGINEERING AIDE: GVW Engineering, Inc., Walnut, CA. Duties included the drafting and design of subdivision infrastructure (tract maps, grading plans, street plans, and storm drain plans), writing legal descriptions, coordinating geometry calculations, and hydrology studies.
June 1992 to present

APPRENTICE SURVEYOR: A-1 Surveying, Inc., Palm Desert, CA. Duties include notetaking, instrument operation, and chaining. Primarily worked in retracement and construction surveying.
June 1991 to June 1992

ACHIEVEMENTS

Passed the Engineer-in-Training examination in October 1994

HONORS AND ACHIEVEMENTS

CHI EPSILON, National Civil Engineering Honor Society, member.

TAU BETA PI, National Engineering Honor Society, member.

WILLDAN ASSOCIATES FELLOWSHIP, nominated by Cal Poly faculty (pending).

ASCE, Student Member

CAL POLY SKI TEAM, races in 91/92, 92/93, and 93/94 seasons. Served as vice president in 93/94 season.

FIGURE 25–2
Résumé—Sample 1

Heading is attractive.

"Graduating" should be deleted.

Throughout the experience section, the description should begin with action verbs such as, "Designed grading plan."

A larger font should be used, continuing on a second page.

MARTHA F. SANDERS
113 Alverson Rd., San Ysidro, CA 92073 (619) 428-1993

EDUCATION
California Institute of Technology, Pasadena
Bachelor of Science in Civil Engineering
Graduating: December 1996
San Diego Mesa Community College
Associate Degree in Psychology
Certificate of General Education

RELEVANT COURSE WORK

Construction Engineering	Reinforced Concrete Design
Highway Engineering	Structural Steel Design
Transportation Engineering	Structural Timber Design
Foundation Engineering	Structural Analysis I & II
Hydraulic Engineering	Statics & Dynamics
Water Supply Engineering	Descriptive Geometry
Technical Engineering	Mech. Drawing
Adv. Surveying	BASIC
Soil Mechanics	FORTRAN

EXPERIENCE
PROJECT FOREMAN ASSISTANT, A.E. Lopez Eng., National City, CA
Duties include grading plan and concrete thoroughfare design coordination and execution. Provide effective liaison between project management and the company. Assist in construction. Provide design practicality advice to assist company estimations. June 1995 to present, June 1994 to Sept. 1994.

CONCRETE CARPENTER ASST., Childress Concrete Const., El Cajon, CA
Duties included interpreting plans and organizing materials to assist formsetters. Interpret instructions to Spanish speaking laborers and assist in construction. June 1993 to Sept. 1993.

ASSISTANT MANAGER, Don Ahles Chevron, North Park, CA
Duties included cashiering 12 automated pumps and generating employee work schedules. Inventory management and computer troubleshooting. May 1991 to Dec. 1992.

CONSTRUCTION WORKER, Haehn Management Co., San Diego, CA
Duties included handling heavy equipment and assisting the on-site inspector to monitor pavement settlement. October 1988 to Jan. 1991.

OTHER SKILLS
Bilingual in Spanish, capable of Hewlett Packard RPN and APPLE CANVAS computer graphics programming.

PROFESSIONAL ASSOCIATIONS
Chi Epsilon, National Civil Engineering Honor Society, council member.
American Society of Civil Engineers, student member.
American Society of Hispanic Scientists and Engineers, student member.

PERSONAL INTERESTS
Restoring classic cars, collecting slide rules, science-fiction, gourmet cooking.

FIGURE 25–3
Résumé—Sample 2

Critique of Résumé—Sample 3

In Sample 3 (see Figure 25-4), the headings are left justified, and the information is included directly opposite them.

Appropriately, this student's GPA includes the basis on which it is calculated, but neglects to indicate if this is a cumulative or major GPA.

The work experience includes details of responsibilities and dates. Each of the duties begins with an action verb. Using two lines for the dates and conditions of employment of the two most recent experiences appears awkward. Also, the street address of the employer should not be included. Sample 3 is attractive and easy to read. It is a well-written résumé.

Critique of Résumé—Sample 4

Sample 4 (see Figure 25–5), unlike the other résumés, was prepared by an engineer with several years of experience and a graduate degree from a prestigious school. Unlike Samples 1 through 3, the information is included on two pages rather than one page. Although one page is recommended, two pages are acceptable when applicable background is sufficient to require two pages. However, each page should indicate that additional pages are included. And these additional pages should indicate that they are a continuation and should include the person's name at the top. This has not been done in this résumé. It is not acceptable to include only a few lines on a subsequent page. Rather, condense or delete information to eliminate the need for this subsequent page.

Notice that in the heading, the writer abbreviates California as "Ca." The accepted abbreviation of California, when included in an address, is CA.

This résumé begins with a summary of qualifications that piques the interest of the reader and directs this reader to the section of greatest interest. In a competitive job market where many résumés may be received for one available position, this prevents the reader from overlooking a desirable qualification.

After either the objective or summary (when included), the résumé ordinarily begins with the most important section of the writer's background. For the recent graduate, unless this graduate has many years of applicable experience, this section is education followed by experience, if any, or relevant courses completed. For a résumé writer with professional experience since receiving the most recent degree, ordinarily, this section is the professional experience followed by education.

This general rule is violated in this résumé, but for a valid reason. Because this writer desires to emphasize completion of a graduate degree at a prestigious school, the section following the summary is education. Further emphasis is placed on this school by including the name of the school before the title of the degree.

The experience section of this résumé appropriately includes quantitative accomplishments without using the proper names of the projects, which have no meaning to most readers.

The information in the other sections of this résumé are included as lists, with each item headed by a dingbat (a graphic font) for emphases.

No doubt, the writer of this résumé will find employment within a short period. The information is easy to find and understand and is professionally presented.

JAMES A. EDEUS, II
905 East Park #56
Carbondale, Illinois 62901
(618) 529–1324

EDUCATION: **Bachelor of Science**
Mechanical Engineering Technology, December 1995
Southern Illinois University at Carbondale

GPA: 3.16/4.0
Deans list three semesters

Associate of Science
Pre-engineering, May 1993
Sauk Valley Community College, Dixon IL

This résumé is attractive and easy to read.

RELATED
COURSES:

Machine Design	Refrigeration	Digital Circuits
Electric Circuits	Thermodynamics	Hydraulics
CAD/CAM	Computer Applications	Robotics

INDUSTRIAL
EXPERIENCE: **National Manufacturing Co.**
1 First Avenue
Sterling, Illinois 61081

The employer's street address should be deleted.

PARTS MANAGER ASSISTANT   June 1995 - August 1995

Inventoried and assigned company part numbers to new and existing repair parts and machine tools, which were contained in the central tool crib.

GENERAL LABORER   June 1994 - August 1994

DIE CAST TECHNICIAN   October 1992 - August 1993
(Part Time and Summer)
Set up and produced various parts on six Techmire zinc die cast machines. Performed quality control. Kept production and maintenance records. Maintained machine operation.

The spacing is awkward here.

PUNCH PRESS OPERATOR   June 1990 - October 1992
(Part Time and Summer)
Produced various products on 100-300 ton Bliss and Minister punch presses using progressive dies. Kept production records. Performed die set-up and maintenance.

PROFESSIONAL
ORGANIZATIONS: Society of Manufacturing Engineers

FIGURE 25–4
Résumé—Sample 3

Use postal codes for state abbreviations in addresses. California should be CA.

JOEL Y. BABIAN
3752 Jones St.
Pasadena, Ca. 91104
(818) 555-3287

SUMMARY

Five years of experience in civil and structural engineering encompassing project design and management, engineering specifications and site inspections.

Design of large-scale water and wastewater treatment facilities including pumping stations, reservoirs, pipelines, storage tanks, and hydraulic structures. Teach CAD at the college level and also conduct CAD training seminars.

Bachelor of Science and Master of Science in Civil Engineering with Professional Engineering registration in Civil Engineering in the State of California.

The summary is excellent.

The information in this résumé is easy to find and clearly presented.

Education

California Institute of Technology
Master of Science in Earthquake/Structural Engineering, June 1992

California State Polytechnic University, Pomona
Bachelor of Science in Civil Engineering, December 1990 - Magna Cum Laude

Registration

Registered as a Civil Engineer in the State of California

Experience

Senior Engineer; Montgomery Watson, Pasadena, Calif; June 1990 to present; responsible for the following projects:

Designed a $4.3 million tertiary facilities upgrade that included a three-story operations building, two 80-foot-diameter clarifiers, and a 12-foot by 16-foot weir diversion structure for inlet control.

Designed and developed piping plan and profile drawings for two 36-inch steel transmission lines. The design transported reclaimed water over a 40-mile stretch of rural terrain. The construction cost was $22.4 million.

Developed earthwork estimates for three international projects in Egypt, Saudi Arabia, and India. The projects were done using the metric system and had grading and drainage plans for 1200 acres of proposed development.

FIGURE 25–5
Résumé—Sample 4 (pp. 403–404)

Computer Proficiency

Macintosh
- Microsoft Word version 5.1
- Microsoft Excel version 5.0
- Macpaint version 2.1
- Filemaker version 2.0

IBM/PC
- MicroStation version 5.0
- Autocad version 12
- Microsoft Windows version 3.1
- Flowmaster Version 1.2

UNIX
- MicroStation version 5.0
- Inroads version 4.0
- Modelview version 3.1
- Design Review version 2.0

Honors and Awards
- Harold Hellwig Fellowship
- Earl C. Anthony Fellowship
- Montgomery Watson Scholarship
- Member of the Deans List (top 1/2 of 1% of college students in America)
- Member of the National Honor Society
- Member of Chi Epsilon

Organizations
- American Society of Civil Engineers

Publications
- *Research Using a Spectrum Analyzer,* 1992

Interests
- Technical Writing
- Athletic Participation
- Music

FIGURE 25–5
(*continued*)

COVER LETTERS

When you send your résumé to a prospective employer, a cover letter is always included. The prospective employer will usually pay more attention to a well-written cover letter than to your résumé; therefore it can greatly enhance your chance of being interviewed. See Chapter 12, "Business Letters and Memos," for a review of business correspondence.

Your cover letter should be on stationery that matches your résumé and should have the same, or similar, heading. It should be individually prepared and signed for each company that you send it. If possible, address it to a specific individual. In your cover letter, you should include the title of the job you are applying for, your professional objective, why you are interested in this company, and details of your background that relate to the position. The closing paragraph should indicate what action you would like the prospective employer to take, or what action you will take.

The cover letter is placed in front of the résumé and attached with a paper clip (when you staple them together, it makes it difficult for the prospective employer to make copies for distribution). A business envelope that matches your stationery adds a complimentary final touch.

You should always keep a copy of the cover letters and résumés that you send, especially when you have more than one standard résumé.

Critique of Sample Cover Letter

The sample cover letter in Figure 25–6 demonstrates the following concerning the contents of the well-written sample cover letter:

- The opening paragraph explains the purpose of this letter.
- The second and third paragraphs discuss the applicant's background as applicable to the position desired.
- The fourth paragraph explains the potential benefit of the applicant's background to the company.
- The closing paragraph tells the reviewer what action is desired.

KENT BROWN
6304 Markham Street
Denver, Colorado 81108
(303) 595–3636

January 5, 1996

Ms. Tami Blake, Manager
Mills Engineering Company
4800 Lakeview Drive
Oakland, CA 94611

Dear Ms. Blake:

I am seeking a full-time position in mechanical engineering with an emphasis in stress analysis. A career in this exciting technology has been my prime motivation for studying mechanical engineering.

I will graduate in June 1996 with a Bachelor of Science in Mechanical Engineering with an emphasis in Machine Technology from Colorado State Polytechnic University. The educational experience at Colorado State includes a strong hands-on approach to engineering problems built on a theoretical foundation.

Since my sophomore year at college I have worked at Rockwell Space Division in Fort Collins. Presently, I design and analyze mechanical components used for rocket boosters.

My work experience combined with my education have provided me with the ability to design and analyze the high-technology projects that are typical of mechanical engineering.

Enclosed is my résumé for your review. I look forward to hearing from you to discuss my qualifications in more detail.

Sincerely,

Kent Brown

Kent Brown

Enclosure

FIGURE 25–6
Sample Cover Letter

THANK-YOU LETTERS

Several days after you have had an interview with a potential employer, sending a thank-you letter will increase your opportunity to receive a job offer.

This thank-you letter may help you in the following ways:

- It will keep your name actively in mind.
- It will let the potential employer know that you are still interested in the company and that you may be receptive to an offer of employment. This would eliminate the time expended by the potential employer with an offer of employment to a noninterested interviewee.
- A well-written letter will demonstrate your ability to communicate.
- It will demonstrate your knowledge of social graces.

In this letter the applicant thanks the interviewer for the opportunity of being interviewed and reminded of the main points of the interview. The closing paragraph should encourage the interviewer to contact you. This thank-you letter is addressed to the specific person with whom you interviewed. Because this will be your latest communication with the interviewer, this letter should demonstrate your interest in the organization and ability to communicate.

Critique of Sample Thank-You Letter

The sample letter in Figure 25–7 demonstrates the following concerning the contents of the well-written thank-you letter:

- The opening paragraph thanks the interviewers, which is the purpose of this letter.
- The second paragraph discusses the potential for the applicant's background to benefit the company.
- The closing paragraph cordially expresses appreciation for the interviewer's consideration and tells what action is desired.

Notice that the sample cover letter and sample thank-you letter are not significantly different in content. However, the sample cover letter contains more details to help persuade the reader to invite the applicant for an interview.

403 Howard Drive
Cleveland, OH 50600

November 11, 1996

Ms. Janet Lee, Director of Engineering
Systems Technology Division
Microwave Electronics
Richmond, IN 54081

Dear Ms. Lee:

Thank you for the opportunity to interview with Microwave Electronics on November 6. Please thank the other members of the interview committee for me.

After discussing your position in electronics systems technology with you, I am very enthusiastic about the possibility of working for Microwave Electronics. I believe my experience as a systems analyst at Farr Company combined with my recent degree in physics would be an asset to your company.

I appreciate the consideration you have given me and look forward to hearing from you. Please call me if you need any additional information.

Sincerely,

Francis Aslind

Francis Aslind
(216) 528–7621

FIGURE 25–7
Sample Thank-You Letter

KEY CONCEPTS

- When changing employment, a résumé is your initial communication with a potential employer. It should be clear, concise, and professional-looking.
- A résumé sent to a prospective employer should always have a cover letter to make a favorable impression.
- A thank-you letter sent after the interview enhances your possibility of being offered employment.

STUDENT ASSIGNMENT

1. Write a résumé for a part-time, summer, or professional job. Search the classified ads for an appropriate position, and write a cover letter to send with your résumé.
2. Discuss the two sample résumés in Figures 25–8 and 25–9. Consider clarity, conciseness, communication of information, and professional appearance. If you were a supervisor and had an opening in your organization, would you invite either of these two applicants for an interview based on the résumé? Why or why not?

STEVEN BAETZ, E.I.T.
98121 Annendale Way, Bellingham, WA 98538–3000
(206) 637–0217

Objective:

Full-time entry-level position in a firm that specializes in structural engineering, or in a structural engineering department of a firm with a variety of specialties.

Education:

Bachelor of Science in Civil Engineering at Washington Technical University
Expected graduation date: June, 1996
Core grade point average: 3.6 (4.0 scale)

Professional Registration:

Engineer-in-Training license number XE090609

Professional Experience:

Intern, Project Engineering department, City of Bellingham, Washington
October 6, 1994 to Present
   Responsibilities: CAD operator. Draft construction drawings for the city. Assist survey crew. Examine storm drains and evaluate for need of repair. Evaluate the arterial and residential road conditions and compute cost estimates of street repairs.

Intern, Manufacturing Engineering department, McDonnell Douglas Aerospace, Monrovia, California
June 28, 1994 to September 22, 1994
   Responsibilities: Write the planning for the production of components that are manufactured by McDonnell Douglas. Maintain computer records of planning.

FIGURE 25–8
Résumé for Student Assignment—Sample 1

Draftsman, Purkiss-Rose-RSI, Bellingham, Washington
September 25, 1990 to June 16, 1994
Responsibilities: Assist the production of architectural and landscape architectural working drawings and presentation illustrations. Compute cost estimates. Computer operation (spreadsheet and word processing).

Leadership Experience:

1996 ASCE Steel Bridge Team Captain for the Pacific Northwest Regional Conference.
1994–95 ASCE Student Chapter President, Washington Tech.
Design Committee Chairman, 1993 Washington Tech's Thanksgiving Day Parade Float.

Computer Experience:

Drafting (AutoCAD, Softdesk Advanced Design and COGO, MicroStation)
Word Processing (Word, WordPerfect, MacWrite)
Spreadsheet (Lotus 123, Excel)
Desktop Publishing (QuarkXPress, Canvas, PageMaker)

Awards:

1994 Bellingham Branch ASCE Scholarship.
Dean's List: Fall 1991, Fall 1993, Winter 1994, Fall 1994, and Fall 1995.

Activities:

ASCE National Student Chapter member 1994 to present.
Washington Tech ASCE Student Chapter member, 1993 to Present.

References Available Upon Request

FIGURE 25–8
(*continued*)

ROSE PANELA
11548 Marcello Way
Marietta, GA 31701
(404) 849–4558

OBJECTIVE:

A design position with a multidiscipline civil engineering company.

EDUCATION:

Southern College of Technology
Bachelor of Science in Civil Engineering
June 1995

RELEVANT COURSE WORK:

- Computers in Civil Engineering
- Computer Programming &
  Numerical Methods
- Technological Economics
- Structural Analysis I & II
- Structural Design of Steel
- Soil Mechanics
- Structural Design of
  Reinforced Concrete
- Hydraulic Engineering
- Water Supply Engineering
- Surveying Computations
- Boundary Control &
  Legal Principles
- Land Survey Description
- Elementary Surveying
- Advanced Surveying
- Geodetic and Satellite Surveying
- Highway Engineering and Design
- Water Quality Engineering
- Engineering Hydrology

COMPUTER SKILLS:

- Microsoft Works
- Windows
- Lotus 1–2–3
- Road/Calc
- Microstation
  Intergraph
- Microsoft Word 6
- XTREE Gold
- Quattro Pro
- Frame 2-D
- Microsoft Power Point
- WordPerfect
- Basic DOS
- Microsoft Excel 5
- Risa 2-D

FIGURE 25–9
Résumé for Student Assignment—Sample 2

HONORS AND ACTIVITIES:

ASCE - 1993, 1994, 1995 Atlantic Southeast Conference, National Student
       Member
CLSA - President (1993–1994), Marietta Student Member
Environmental Engineering
Engineering Council Member - 1993, 1994
Tau Beta Pi
Chi Epsilon - 1995–96 Editor
National Science Scholar

WORK HISTORY:

MICROSTATION LAB TECHNICIAN
        Monitor computer stations and assist students with difficulties.
        Southern College of Technology
        October 1994 to Present

PD&C - SURVEYS: STUDENT ENGINEER
        Operated instruments of measurements (G. P. S., E. D. M.,
            Thermodolites, and Levels).
        Input constructed transmission lines into the G. I. S.
        Atlanta Department of Water and Power
        June 1993 to September 1993
STUDENT ASSISTANT
        Checked incoming digitized maps and prepared them for approval
            and finalization.
        Checked aerial photos and layout flight plans.
        Created local data base for To-Reaches and graphic files for USGS
            maps
        Marietta, Georgia
        November 1993 to January 1994
        June 1992 to September 1992

FIGURE 25–9
(*continued*)

Index